Imagining Europe

Europe and European Civilisation as Seen from its Margins and by the Rest of the World, in the Nineteenth and Twentieth Centuries

P.I.E. Peter Lang

Bruxelles · Bern · Berlin · Frankfurt am Main · New York · Oxford · Wien

Michael WINTLE (ed.)

Imagining Europe

Europe and European Civilisation as Seen from its Margins and by the Rest of the World, in the Nineteenth and Twentieth Centuries

"Multiple Europes"
No. 42

Cover picture: Daniel Chester French, The Continents Personified: Europe, America, Africa. Bowling Green, New York City, 1907.

Éditions scientifiques internationales
Brussels, 2008
1 avenue Maurice, B-1050 Brussels, Belgium
info@peterlang.com; www.peterlang.com

ISSN 1376-0904
ISBN 978-90-5201-431-9
D/2008/5678/39

Library of Congress Cataloging-in-Publication Data
Imagining Europe : Europe and European civilisation as seen from its margins and by the rest of the world, in the nineteenth and twentieth centuries / Michael Wintle (ed.).
p. cm. — (Multiple Europes, ISSN 1376-0904 ; no. 42)
Includes bibliographical references and index.
ISBN 978-90-5201-431-9
1. Europe—Civilization—19th century—Public opinion. 2. Europe—Civilization—20th century—Public opinion. 3. Europe—Foreign public opinion. 4. Europe—Relations. I. Wintle, Michael J.
CB417.I43 2008 940.2'8—dc22 2008033670

CIP also available from the British Library, GB.

Bibliographic information published by "Die Deutsche Bibliothek". "Die Deutsche Bibliothek" lists this publication in the "Deutsche National-bibliografie": detailed bibliographic data is available on the Internet at <http://dnb.ddb.de>.

Contents

List of Illustrations

Preface

This book, in its present form, is one of the eventual products of a workshop which took place at the Royal Academy in Amsterdam in June 2005, to honour the retirement of Bruno Naarden as the Professor of Russian History and East European Studies at the University of Amsterdam. At that well attended and lively symposium, which included Professor Naarden's valedictory lecture, on the subject of the Amsterdam *burgemeester* Nicolaas Witsen's writings on Russia in the seventeenth century, four of the contributors to this book gave embryonic versions of their chapters as papers to the workshop. Since then they have been reconsidered and entirely rewritten, in order to address more directly the common research agenda; moreover five entirely new chapters have been specially commissioned from other specialist scholars. We have moved well beyond Bruno Naarden's farewell celebrations, but nonetheless this book is dedicated to him as a token of our appreciation for his untiring work for European Studies at the University of Amsterdam and elsewhere.

I am grateful to those who helped plan the original occasion, especially Ben de Jong, and to all those who took part in the workshop. I am also happy to thank the Institute for Culture and History at the University of Amsterdam, which provided generous financial support for the project at various stages. Most of all I am indebted to the authors, who have been infinitely patient and understanding throughout the vagaries of the editing and publishing process: both their scholarship and forbearance are much appreciated.

Michael Wintle

Amsterdam, December 2007

PART I

INTRODUCTORY

INTRODUCTION

Perceptions of Europe within and without

Michael WINTLE

This collection of essays revolves around the highly topical question, "What is Europe?" What are its values, what are its borders, and what does it stand for? Many academic enquiries have posed these questions; this book, however, asks them from a different point of view.

Perception and image-formation often have to do with identity formation: images of self-perception are what we use to define our own identity, and we are helped in doing so by our perceptions or images of what we are not: our opposites, or Others. Europeans too have formed a collective identity in the same way, certainly since the Renaissance, defining what unites them, and what they think they are not. But what have the others thought of Europe? The authors of the essays in this volume are thus not so much interested in what Europe thinks of itself, but rather in what others think of it. The contributors take a number of scenarios from recent history, and examine how Europe has appeared to people in other parts of the globe: America, China, the Arab world, for example. But we go further. The question is also posed for some parts of the world which are "inside" Europe, but which for one reason or another hover on the margins, like the Balkans, and Turkey. Also included are the views about Europe held in parts of the continent which have without any doubt whatsoever belonged to Europe's core, but which much of the rest of Europe, later, would like to forget about, or marginalise: Stalin's Russia, and Hitler's Germany.

Among these outlook positions with their views on Europe, some are conventional external Others, like those which are indisputably outside geographical Europe; some are various forms of internal Others. Of the examples chosen here, perhaps only China represents a complete alterity for Europe, and *vice versa*. All the rest share something with Europe – often something crucial and even essential – through post-colonial hybridity in Latin America, the jointly held "Western" civilisation in the case of the US, the geographical location of the Balkans in southeast Europe, and the centrality of totalitarianism to European "civilisation" in the case of Stalin and Hitler. Nonetheless, these views of Europe all differ from those *self*-views conventionally held by the modern democracies of the EU, which see (and like to see) great differences between themselves and the parties dealt with in this volume. However, most of the elements elaborated here are central to the imagining of Europe, and despite many Europeans' wish to distance themselves, such views should be recognised and taken up as an important and indispensable contribution to the debate about "What is Europe?"

The approach taken here is multi- and interdisciplinary, with a focus on political and cultural history. The evidence used to answer the question of what Europe has meant to the world varies from Stalin's speeches and Mustafa Kemal's writings to literary texts, paintings and maps. The theoretical approach takes account of concepts of identity, Othering, geopolitics and the psychology of international relations. It complements well the large number of publications on what Europeans think about Europe, and puts specialised views of Europe and/or the West, like Kagan's *Paradise and Power*, or Buruma and Margalit's *Occidentalism*,[1] into a global and comparative perspective.

The collection starts with a broad and so necessarily selective survey by Michael Wintle of views of Europe "from the outside". This introductory chapter uses as evidence mainly the *visual* images which non-Europeans have held of Europe. This contribution to "visual history", then, literally looks at various images of Europe (in the sense of visualisations), for example in Arabic and Asian maps, colonial and postcolonial paintings, and American cartoons, drawings and statues. The conclusions drawn are that representations of Others are indeed crucial to identity formation, that Europe was quite often ignored by the rest of the world in the past, particularly before the nineteenth century, and that in the modern period the relationship with Europe was almost invariably complex, intertwined and hybridised.

Following this introductory section, Parts II and III deal with views of Europe from what might be called, for quite different reasons, its "margins". The first category consists of the views of totalitarian Europe, specifically Stalin's views of "European civilisation", and the Nazi vision of Europe. The second category, in Part III, is marginal in the sense of being on the geographical edge of Europe: Turkey (especially under Atatürk), and the Balkans. They are clearly part of the continent, but at various times actively excluded from "Europe" by other "Europeans". These are border regions, in the sense that they glide in

---

1 R. Kagan, *Paradise and Power: America and Europe in the New World Order*, London, Atlantic, 2003; I. Buruma & A. Margalit, *Occidentalism: a Short History of Anti-Westernism*, London, Atlantic, 2004.

and out of Europe, subject to the prevailing political and cultural situation. The totalitarian regimes in Part II certainly belong to the European core, like the fascist ones in the south, but at the same time they have also been banished by other Europeans to (at best) a periphery. Perhaps the most interesting case is that of Nazi Germany. There can be no doubt whatsoever that the Nazi experience is absolutely central to the experience of Europe. Unpalatable as it is, the Third *Reich* and the Holocaust are as essential to European identity as are Charlemagne and the EU. It indicates adroitly that identity politics are actually much more complex than "them" and "us", for there are internal and marginal Others which are just as important to self-definition as external or constituting Others. It has been argued that the Nazi episode (itself a marginalising expression) functioned for a long time as Europe's Other,[2] and perhaps it still does, but it constitutes at the same time an essential part of what Europe was and is.

In Chapter 2, Erik van Ree treats Joseph Stalin's views of the various national characters in Europe, which he classified on the basis of the archetypes of the hero and the merchant. The Soviet leader divided the nations of Europe, roughly, into those of the East and those of the West, measuring them along two axes of civilisation and strength of character, which were more or less negatively correlated. The Eastern nations he believed to be relatively uncivilised but endowed with heroic, martial qualities. The Western ones were considered the more developed, but with their commercialist orientation they suffered from a deplorable lack of iron in their character, in Stalin's view. Van Ree also explores the remarkable similarities of Stalin's image of Europe with that also nurtured by German Conservative Revolutionaries such as Werner Sombart.

David Barnouw takes on the difficult subject of the Nazis and Europe in Chapter 3. After the defeat in World War I, Germany felt itself crippled and asset-stripped: the Treaty of Versailles had relieved it of more than ten per cent of its territory. Most Germans wanted to regain at least some parts of those losses, but the Nazis in particular wanted a complete *Neuordnung* of Europe, to make Germany as large and strong as possible and to prevent another humiliation like Versailles. This hunger for land, or for *Lebensraum*, required a scientific grounding. Hundreds of scholars were involved, harnessing their academic skills to the promotion of an expansionist and often racist policy. Geopolitics, *Ostforschung* and *Westforschung* were discourses which paved the way to the conquest of Europe, in order to give Germany and the Germanic

2 E.g. in T. Lawson, "The Myth of the European Civil War", in R. Littlejohns & S. Soncini (eds.), *Myths of Europe*, Amsterdam, Rodopi, 2007, pp. 275-289.

people their historic place as masters over inferior peoples. Barnouw offers a critical survey of the different scholarly currents involved in the shaping of a new, Nazi view of Europe, and some analysis of the fortunes of those scientific activities since the War.

In Part III, on the Margins of Europe, we begin with Turkey's view of Europe, which is deeply ambivalent, and a good example of a marginal, changing and divided position. As Erik Zürcher shows in Chapter 4, on the one hand Europe (and especially Western Europe) has been seen as an example to emulate, if Turkey were to become truly strong as well as modern. The Turkish upper and middle classes, with their strong commitment to secularism, define the world in black and white terms. In their eyes, being a part of Europe (or more generally, the West) is imperative if Turkey is to be secure against attempts to reintroduce an Islamic social and political order. In the eyes of these Kemalists, modernity, civilisation and the West are one. On the other hand, Turkish nationalists, whether religiously inspired or not, see Europe as the force that carved up the Ottoman Empire and tried to put an end to the very existence of Turkey as an independent state. They suspect that Europe, or the West, has a role in every attempt to undermine the Turkish nation state, whether it originates among the Armenian diaspora or the Kurdish minority. Phrases like "a new Sèvres" and "crusader mentality" abound in the popular media, and spread an image of Turkey under permanent threat. Both images, that of Europe as an example and that of Europe as a threat, go back to the nineteenth century and are deeply embedded in the Turkish psyche.

Secondly in this section on the geographical "margins" of Europe, Joep Leerssen engages with the issue of the Balkans, in Chapter 5. Much outstanding work has been done in recent years on the European view (and indeed construction) of the Balkans.[3] However, the process also worked the other way: there is such a thing as a Balkan construction of Europe, and Leerssen's chapter examines that process. When the European ideology of Romantic nationalism hit Ottoman Europe, intellectuals and activists from various Balkan regions conceived, and worked on the basis of, a specific idea of Europe. This served both as a point of identification (an alternative to the Ottoman Empire) and as an uneasy semi-Other (as the locus of Western Christianity). To chart this process, a variety of very different language traditions must be taken into account: Greek, Albanian, Romanian, Serb, Bulgarian. The intellectual network that spanned nineteenth-century south-eastern Europe

---

3 One example among the many: L. Wolff, *Inventing Eastern Europe: the Map of Civilization on the Mind of the Enlightenment*, Stanford, Stanford UP, 1994.

worked across these linguistic divides, however, and ideas on Europe travelled freely and dynamically between Vienna, Odessa and Istanbul.

In Part IV we move to consider the view of Europe from areas certainly geographically defined as being outside Europe, and which generally considered themselves extra-European as well, albeit with specific links, positive and negative, with the European continent or civilisation. In Chapter 6 Ruud Janssens investigates the views of Europe held by Americans in the United States. To Americans, Europe and Europeans have been at various times and simultaneously a threat, a trade opportunity, and a cultural ideal. In the context of national security, it was George Washington, the first President of the United States, who warned his countrymen against entangling alliances with Europeans. In his view, Europeans always fought wars among themselves, and he wanted Americans to avoid getting caught up in such conflicts. Trade with Europeans was fine, but there were to be no military alliances. This perception of Europeans has essentially remained in place. The main difference is that since the end of World War II, American political leaders have decided that Europeans cannot be left to their own devices, because of their track record in involving the Americans in major conflicts. Consequently, American governments supported European integration and NATO at the end of the 1940s, and their expansion during the 1990s. Cultural relations with Europe were at least as complicated for Americans as the issue of national security. Especially in the nineteenth and early twentieth centuries, Europe was an ideal of civilisation for Americans, manifested in the presence of many wealthy Americans in Europe, American academics at European universities, and American artists in cities like London and Paris. Yet it was also important to distance themselves from Europe, to show that the United States created its own culture, and was not just a lesser version of European culture. In this chapter these different approaches to Europe by Americans are brought together. It is intriguing to note that, although Americans are very aware of ethnicity in their own country, they tend to think more in terms of a "European culture" and "Europe" than Europeans, who often consider themselves in more nationalistic terms. Janssens finishes with an examination of several Hollywood filmmakers' views of Europe, which invariably embodied a juxtaposition of contradictory feelings, such as superiority and longing: Europe was a place that it was necessary to escape both *from* and *to* at the same time.

In Chapter 7, Stefan Landsberger casts his eye over Chinese views of the West and of Europe, and does so by using the prism of images in popular culture, not least his own collection, unrivalled in the West, of Chinese propaganda posters. In this book it has not always been easy to separate ideas about Europe from ideas about the West in general. In the

American (US) case there is clearly an ambivalence, because Europe and the US are both prime players in "the West"; in the case of China, however, the Chinese consider themselves entirely outside both Europe and the West. "The West" as a concept does not have a very long pedigree, dating really only from the Second World War, and it may well be falling apart now, with the collapse of the Soviet Union and increasing divergence between the US and the EU in key policy areas. In many fields, like development economics, North/South is a now more meaningful watershed than East/West. Nonetheless, Landsberger's chapter covers the period since the end of World War II, and he has to tread carefully along the fine line between the two concepts of "West" and "Europe". After the People's Republic opened its doors to the outside world in the late 1970s, living conditions improved and ownership of radios and televisions became widespread. As a result, images of events and peoples outside China started flooding in. In previous decades, the appearance of people from the West in the media had been crafted carefully and determined largely by China's identification of friends and foes. Propaganda posters played an important role in the construction of these images of the Other, supplementing newsreels and press photographs. To complicate matters further, another categorisation enters the scene: that of race. When analysing the images of non-Chinese in the Chinese media, it is impossible to avoid reference to Chinese feelings of ethnic superiority. They resulted in the Chinese traditional ordering of mankind into a racial hierarchy of biological groups, led by the yellow and white races. As a consequence, there was often no real distinction made between the origins of the Westerners shown: they were all just white. After the founding of the People's Republic in 1949, political allegiances became much more important. Images of foreigners almost disappeared by the early 1970s, when China had few friends left in the Western world. Finally, Landsberger examines the carefully constructed image of the West that has come to dominate television in present-day China.

In Chapter 8, Robbert Woltering tackles the diverse subject of the Arab view of Europe, in which he finds almost infinite variety. There is of course at this time every need to improve our understanding of the relations between the Western and the Arab worlds, for the Arab world is the Muslim region nearest to Europe. Many recent publications have tended to focus on past and present animosity between the northern and southern shores of the Mediterranean. Woltering takes a much wider look at Arab perceptions of Europe by giving due credit to the modernisation project of the *nahda* (Arab renaissance), and Islamic reforms of the nineteenth and early twentieth centuries, which came with a sometimes ambiguous but generally positive image of Europe. These findings are then used in the subsequent analysis of appreciations of Europe in post-Cold War cultural and political discourse among Arab intellectuals, mainly from Egypt. His account also pays attention to the way in which Europe and the EU are contrasted with the United States in the Arab media.

Finally, Peter Beardsell moves in Chapter 9 towards establishing what Latin American views amount to, and a very complex matter it is too. The colonial experience caused a legacy of mixed images of Europe in Latin America. While many there regard themselves as the direct descendants of the original indigenous civilisations, there are many others who trace their descent straight from Europe. The profound effects of racial and cultural hybridity add to the complexity, as do social and political factors. Latin America's relations with the United States play their part too. For these reasons Beardsell analyses contrasting images of Europe: as invader, neo-imperialist power, destroyer of the natural world, an alien culture; and on the other hand, as a model of civilisation in opposition to barbarity. He then addresses the issue of Latin America's identity, in which Europe has a fundamental role as both Self and Other.

This collection of studies concerning what some people outside what might be called "mainstream Europe" have thought about "Europe" in various times and various places is of course not systematic enough to be able to deliver an unequivocal answer to the question of "What is Europe?", whether in a geographical, political or cultural sense. Nonetheless, some provisional conclusions are possible, and need to be drawn. In the first place, not everyone thinks about Europe at all: often it was entirely ignored, especially before 1800, and by such powerful parts of the world as China for long afterwards. However, in the last two centuries Europe has usually generated a response of some sort, and the feelings are mixed. There has been, of course, resentment at the overbearance of European economic, political and cultural imperialism, but there is also respect for aspects of European culture and sophistication. We have seen disgust for the tyranny of commercialism and decadence which has been held to engulf Europe, as well as cultural legitimation. The image of Europe has been caught up, often rather inextricably, with the concept of "the West" in the twentieth century; this complicates matters, especially for China, but on the other hand some parts of the world do indeed draw a close distinction between Europe and the West, and where they do not it is yet further evidence that Europe does not always evoke any reaction at all, however strange that may sound to Europeans. That itself is a conclusion worth drawing. But the principal finding of this rich set of studies must be that the Othering process, as part of identity formation both in Europe and all other parts of world, is unendingly complex in its workings, and defies simplistic characterisation in terms only of "us and them". Hybridity is the kind of concept which indicates that complexity, and Peter Beardsell's implicit allegiance in Chapter 9 to the theoretical framework offered by Homi Bhabha may offer lessons to us all. At the very least we should recognise that absolute rejection or absolute adulation of any Other, external, internal or intermediate, are hardly ever the case in reality, now or in the past, despite the strident voices of so many in the political and media worlds.

CHAPTER 1

Europe as Seen from the Outside

A Brief Visual Survey

Michael WINTLE

The imagining of Europe has been a much-visited subject in recent years, as the European Union struggles to define itself and its role in the world. There are Euro-enthusiasts who would like to expand or at least strengthen the supranational structures and powers of the Union, and who believe that the way to legitimate that course of action is to define and develop a European identity, akin to the way in which a national identity legitimises a national government and state apparatus. There are just as many who disagree with that project fundamentally, and who set out to prove that there is not and cannot be a European identity, or at the least that it is quite different from national identity. All this stimulates the academic interest in what Europe is, has been, will be or should be: how to define Europe is now a constantly posed question, and it has been a topic of periodically intense interest since the Renaissance. Of interest, that is, to Europeans.[1]

For nearly all of these definitions, imaginings, views and images of Europe have been self-images. Moreover, most of these images have been concerned to glorify Europe in one way or another; they are Eurocentric. When looking at *visual* images (the main burden of this introductory chapter), maps have been very important, and the great example of a Eurocentric cartographic image of the world is of course the Mercator projection, dating from 1569. That view of the world, emphasising the centrality of Europe and exaggerating its relative size, eventually became universally familiar amongst Europeans, and the great majority of European maps of the world ever since have opted for the Eurocentric Mercator projection.

1 A sample of relevant studies: G. Delanty, *Inventing Europe: Idea, Identity, Reality*, Basingstoke, Macmillan, 1995; M.J. Wintle (ed.), *Culture and Identity in Europe*, Aldershot, Avebury Press, 1996; B. Stråth (ed.), *Europe and the Other and Europe as the Other*, Brussels, PIE Peter Lang, 2000; R. Littlejohns & S. Soncini (eds.), *Myths of Europe*, Amsterdam, Rodopi, 2007.

Figure 1.1. Gerard van Keulen, World Map, "Paskaart vertonende alle bekende zeekusten en landen op den geheelen aardboodem of werelt", Amsterdam, c. 1720

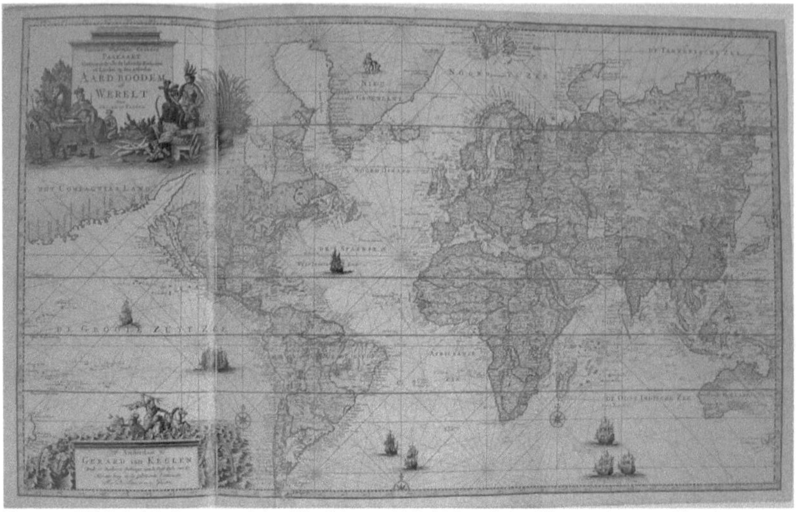

Source: Whitfield, *Image of the World*, pp. 108-109; original in a private collection.

A fine historical illustration is Gerard van Keulen's world map of about 1720, reproduced in Figure 1.1. Van Keulen came from a cartographic family of long standing; he was official hydrographer to the mighty Dutch East India Company (VOC) at the height of its power, and his maps "enjoyed an unrivalled position of authority during the eighteenth century".[2] The map bristles with rhumb lines and geographical data: it is conducting a hegemonic discourse of authoritative spatial ordering. It covers a full 360 degrees of longitude, and apart from the serious compression of Asia, the assertion that California is an island, and some understandable vagueness about the North American west coast, it is remarkably accurate in outline, especially in South-East Asia, where the Dutch colonial interest predominantly lay. A number of features stand out. First, it is a colonial map: Van Keulen was in the service of the VOC, so there are copious references to "the Company", Dutch-language labels cover the sea coasts, and Dutch East Indiamen dominate the seas. Second, it uses wide open spaces, especially in America and Australia, but also in parts of Africa and Asia, to indicate vacuity and vacantness. Third, it is a hugely Eurocentric map in a very literal sense: there is hardly a better example to be found of the Mercator

[2] P. Whitfield, *The Image of the World: 20 Centuries of World Maps*, London, British Library, 1994, p. 108.

projection, grossly enlarging the size and importance of Europe on the world map and proportionately reducing the size of the other continents, especially Africa and South America. Europe is longitudinally plumb in the centre of the map, because the chart has been broken in the Pacific (and Asia foreshortened), but also in latitude terms, for the bottom of the map has been removed, which places Europe in the prime position. It is in the central, controlling location.

Furthermore, Van Keulen's map embodies the imperialising function of cartography in the eighteenth century in its marginal decorations. The lower cartouche, on the left, consists of self-promotion for the family's map and book business in Amsterdam. The upper cartouche, however, is a particularly fine example of the colonial relationship in a personification of the four continents, and is a remarkable statement of Europe-assertion. It is a highly detailed rendering, based on traditional iconography, and is an eloquent manifestation of the way in which geographical iconography, as well as the maps themselves, can reinforce the colonial relationship. Europe is graciously receiving the tribute of her sister continents, including Asia, who carries a crescent sceptre for Islam, and who has an incense-burner at her feet: she offers Europe a casket of jewels and a gold ornament. Primitive black Africa and Red Indian America have brought ivory, silver and trade goods, while they are accompanied by a snake and a lion. Europe meanwhile sits at a table, reading a book which speaks for the reason for her superiority: knowledge and wisdom. She wears the trappings of power, and a lyre at her feet alludes to the arts. Beside her kneels a servant with the true church. Meanwhile, should there be any residual doubt whatsoever about the precise nature of this continental hierarchy, a rather fierce-looking white warhorse, the emblem of Europe's supremacy in arms and warfare, lurks in the bushes behind to make sure that the proprieties of rank are adhered to. This is indeed, together with the cartographic features of the image as a whole, an extremely compact expression of imperialism, and of European views of global geopolitics in the age of empire.[3]

This, and thousands like it, are highly Eurocentric visual images of Europe fashioned by Europeans, and they are fairly typical images of the world held by Europeans since the Renaissance. To put oneself at the centre of one's world may be childish but it is also human, and certainly not exclusive to Europeans. In that sense there can have been no more self-obsessed people than the Chinese, for example, who have always put their "Middle Kingdom" at the centre of everything.[4] Eurocentrism

---

3 See also Wintle, *Culture and Identity in Europe*, pp. 84-85.

4 J.B. Harley & D. Woodward (eds.), *The History of Cartography*, Vol. II, Book II, *Cartography in the Traditional East and Southeast Asian Societies*, Chicago, Uni-

has dominated the European vision of the world for the last five hundred years and more, and perhaps that is hardly surprising.

These visual images of Europe form a very powerful part of the identity formation process. Perception and self-perception underpin identity, and European identity is an expression of self-perception and self-images, including the visual ones. These images inform us about European identity, in the form of differences perceived by Europeans between themselves and their Others, especially in regard to ideas about civilisation and its opposites, and an ideology of Eurocentrism. All these things interact with the creation of images of Europe.

But again, these images are by Europeans: they are the ones most interested in Europe, after all. This volume, on the other hand, is concerned with the images of Europe held by people outside Europe: by some of those on the margins of Europe, like Stalin and Atatürk, and by those whom Europe has since *tried* to marginalise: the Nazis, and the inhabitants of the Balkans. These images of Europe are not the standard, Eurocentric self-images of Europeans in past and present.[5] In this chapter, by way of introducing the other essays, we shall take a brief look at what the rest of the world has thought of Europe, as opposed to what Europe has thought of its own position in the world. The subject is therefore Europe from the outside, or from the *other* side, and will examine some of the *visual* images of Europe created by non-Europeans. The selection will be arbitrary, for this author possesses neither the resources nor the linguistic skills for a systematic survey. Most of the work on European identity is concerned with "the European gaze" directed at its Others, but some kind of account, however fragmentary, of the "returning gaze", of the Others looking in, is also essential. In concentrating on the visual aspects of that representation, it is intended to form a suitable introduction to the other chapters in this collection, which deal in a more specialist and comprehensive way with views of Europe from outside its borders, or from its margins. In the words of Luisa Passerini, a fine scholar of Europe, "Let us [also] accept the representations and images of Europe produced *outside* of Europe".[6]

---

versity of Chicago Press, 1994, p. 173. See also the chapter in this volume by Stefan Landsberger, and the section on Chinese maps in the present chapter.

5 See my forthcoming study, M.J. Wintle, *The Image of Europe: Visualizing Europe in Cartography and Iconography*, Cambridge, Cambridge UP, 2008 forthcoming.

6 L. Passerini, *Figures d'Europe: Images and Myths of Europe*, Brussels, PIE Peter Lang, 2003, p. 32 (my italics).

Cartography

When it comes to visual images of parts of the world, then maps are of central importance. Now maps are not unbiased documents in this respect. There is no such thing as an objective map; a map always has an agenda (conscious or not); a map is always a product of data selection and representation, operating on the basis of a set of assumptions which are cultural and political. We have already noted the example of the Mercator projection as a mapping technique of excellent scientific qualities, but which was and is also profoundly Eurocentric. Maps reflect and indeed promote the cultural assumptions and aspirations of their makers and patrons: so much is now generally accepted by historians of cartography.[7]

Figure 1.2. "The New World Map, Wizard's Projection", Christchurch, New Zealand, c. 1990. By permission

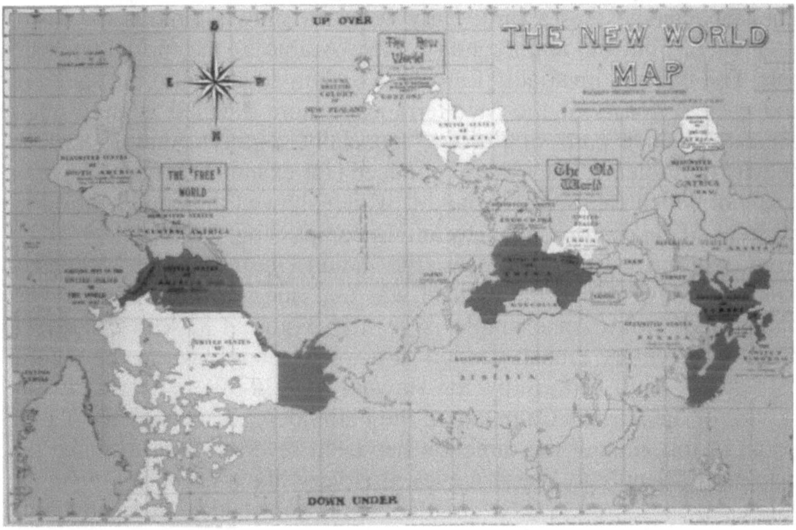

There are several ironical inverted Mercator projection maps in circulation nowadays, many of them originating from Australasia. One example is the "The New World Map, Wizard's Projection", produced by Ian Brackenbury Channell, who was employed by the public authorities in Christchurch, New Zealand, as a kind of tongue-in-cheek, semi-official soap-box speaker, from 1980 to 2002 (Figure 1.2). South is at

---

7 M.H. Edney, *The Origins and Development of J.B. Harley's Cartographic Theories*, Monograph 54, *Cartographica*, 40/1-2 (2005), Toronto, University of Toronto Press, 2005; and Wintle, *The Image of Europe*, Chapter 1.

the top (and why should it not be?), while Europeans are relegated to the extremities of the map usually reserved for their Antipodes. Indeed, the whole notion of continents is ridiculed, while at the same time showing how potent an implement of Europe-assertion the Mercator projection has been for the last 450 years.

Over the centuries, most maps of Europe have been drawn by Europeans, but some have also been produced by cartographers working outside the European traditions. The main categories are Islamic or Arabic maps, and Asian maps, from India and the Far East. In both cases we are fortunate in being able to rely upon existing, excellent surveys in the standard work on the subject, the University of Chicago Press's *History of Cartography*, edited (in its earlier volumes) by Brian Harley and David Woodward.[8]

From the seventh century onwards, a great Arab/Islamic cartographic tradition slowly grew up. One charming feature was to portray the Eurasian landmass as a bird: early versions show China as the head; later ones, from the tenth century, have Arabia as the most important part. The tail was invariably Europe. As Islam advanced to the southern and eastern coasts of the Mediterranean, more detailed Islamic maps came to include parts of Europe, especially southern Spain. Drawing heavily on Ptolemy, maps were produced (for example by Al-Istakhrí, or a school which followed Al-Balkhí) which showed the known world as a circle of land surrounded by an Encompassing Ocean and pierced from east and west by two seas (the Indian Ocean and the Mediterranean), south of which the climate was hot, and north of which it was increasingly cold. An example is seen in Figure 1.3, which shows a nineteenth-century French version of a world map drawn in the second half of the tenth century by Ibn-Hawqal (c. 930-90), who was in the service of the African Caliphate of the Fatimides. He would have placed South at top, though the direction has been reversed here to make the map more friendly to Western eyes. It is very like a European medieval T-O map in some ways: a circular world surrounded by an external ocean and split into continents by sea-arms and rivers. Africa (*Pays des Noirs*) is shown to the left, or West; most of the map is concerned with the Muslim areas of Asia and especially the Arabian peninsula. Europe is a small, triangular island, mentioning Andalus; the Mediterranean islands of Cyprus, Crete and Malta are plain to see. Europe featured on these maps as a periphery, as was to be expected. There was simply not

8 J.B. Harley & D. Woodward (eds.), *The History of Cartography*, Vol. II, Book I, *Cartography in the Traditional Islamic and South Asian Societies*, Chicago, University of Chicago Press, 1992; and Harley & Woodward (eds.), *The History of Cartography*, Vol. II, Book II.

very much interest, and not very much data, on the part of these early Islamic cartographers, about that remote corner of the world.[9]

Figure 1.3. Ibn-Hawqal, World Map, c. 980

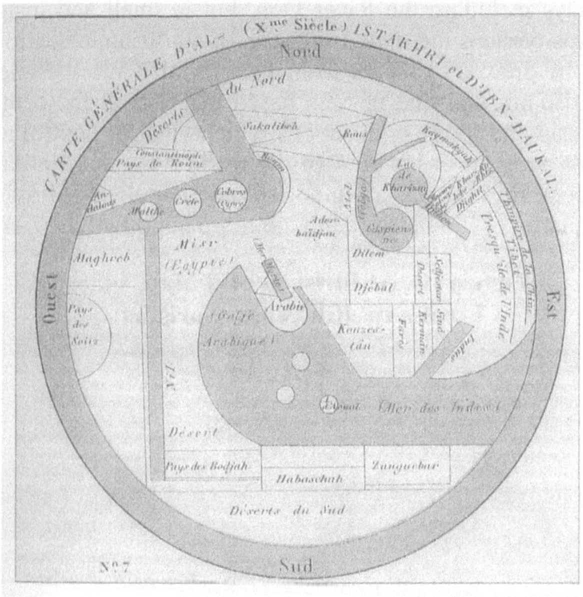

Source: V. de Saint-Martin, *Atlas dressé pour l'histoire de la géographie*, Paris, Hachette, 1874.

In the twelfth century, the great mapmaker Al-Sharíf al-Idrísí was active, travelling widely from his birthplace in Ceuta (as far as England, for example) before entering the service of the Norman King Roger II in Palermo. His work was also based heavily on Ptolemy and, because of his Norman patronage, his cartography expressed much more interest in Europe than had previous Islamic maps. A magnificent world chart in rectangular panels made for King Roger in about 1150 AD showed Europe in great detail and accuracy, beyond anything that Europeans were producing at the time. He also made a single, circular world map as an accompaniment, shown in Figure 1.4, which looks remarkably like a medieval *mappa mundi*, a tradition on which it drew. South was at the top, and it is not shown inverted here, as was Figure 1.3; the religious

9 Harley & Woodward (eds.), *The History of Cartography*, Vol. II, Book I, pp. 90-91, 118-123; J.G. Leithäuser, *Mappae Mundi: die geistige Eroberung der Welt*, Berlin, Safari [1958], pp. 103-113; P.H. Gommers, *Europe – What's in a Name?*, Leuven, Leuven UP, 2001, pp. 19-21.

content which was the essence of medieval Christian maps is absent, except for the fact that the Arabian peninsula is in the centre of the map. Africa shows the Nile and its source; the continent is seen to be vast. Europe is clearly recognisable in the bottom right-hand corner, bounded by waterways including the River Don, but is small and insignificant. Al-Idrísí was perhaps the most European of all "Islamic" cartographers, given that he drew on sources from East (Arab), West (Ptolemy) and North (Christian), and that he was based in Sicily at the centre of the Mediterranean and of the civilisations which vied for control around its shores. As a geographer his view of the world was quite similar to Ptolemy's; Europe for him was an important and populous part of the word – about a quarter of it – but certainly no more important than any other.

Figure 1.4. Al-Idrisi, World Map, 1456.
From a twelfth-century original

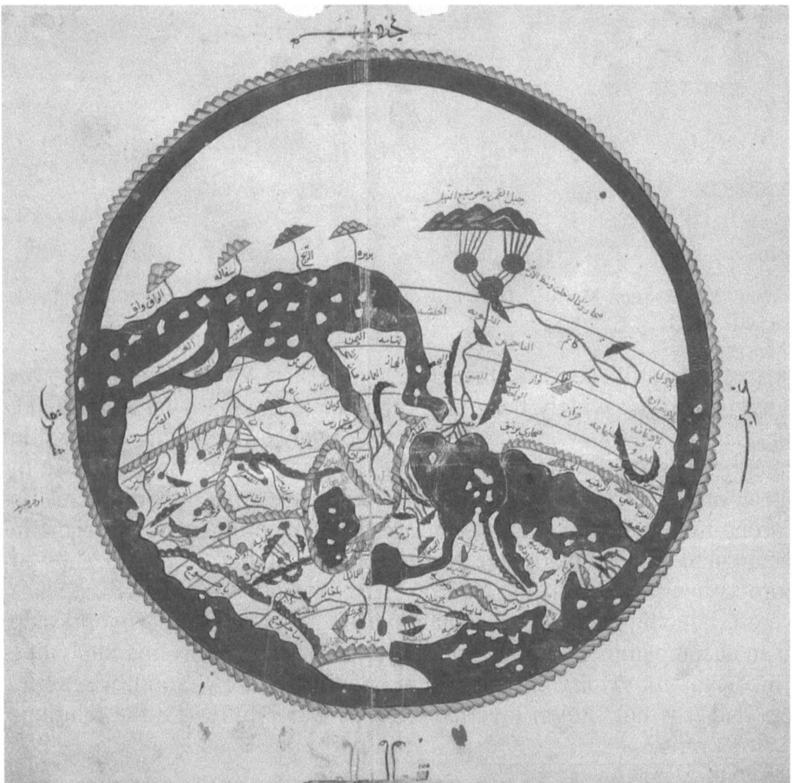

Source: Whitfield: *The Image of the World*, p. 29.

Figure 1.5. Map of the world by Al-Qazwini, 1388

Source: C. Hillenbrand, *The Crusades: Islamic Perspectives*, Edinburgh, Edinburgh UP, 1999, plate 6 (original in Bibliothèque Nationale de France).

Some later Arab maps did indeed concentrate on religion, marking the Muslim pilgrimage sites as the centre of attention. One map of the world by Al-Qazwini of 1388, shown in Figure 1.5, was highly schematic, and is hard for Westerners to read if they do not have the cultural equipment to decode and understand the conventions. Again, South is at the top. The round earth is set in an ocean surrounded by a pasture. Water inlets to the world are clearly marked, including the Mediterranean and the Gulf, and three special sites are shown in yellow, probably the principal pilgrimage destinations.

Figure 1.6. Islamic Chart of Europe, the Near East and North Africa, mid-sixteenth century

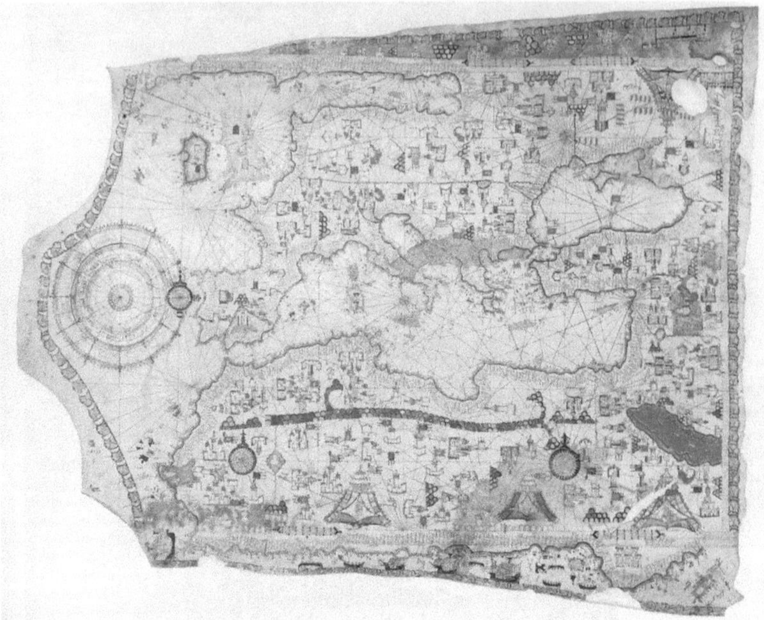

Source: *ESF Communications*, 44 (Spring 2002), p. 21 (original in Topkapi Sarayi Muzesi Kutuphanesi, Istanbul (ref. H.1822)).

Later, Ottoman maps of the period from the sixteenth century onwards draw equally on these Islamic and European traditions.[10] To complete this canter through some aspects of the early Islamic cartographic picture of Europe, Figure 1.6 shows a chart of the continent by Hajj Abu al-Hassan, dating from the mid-sixteenth century and now in the Topkapi Museum in Istanbul. The coastlines of the Mediterranean and the Black Seas are highly accurate, whereas the Atlantic coast, especially in the North, is vague. A mass of detail is depicted in Africa and the East, whereas Western maps at that time would have shown those areas as blank. In general, before the end of the eleventh century "the Muslims knew little and cared less about Europe; it just did not impinge much on their world view". Their geographic knowledge of

10 Whitfield, *The Image of the World*, pp. 28-29; L. Bugner (ed.), *The Image of the Black in Western Art*, 2 Vols (in 3 parts), Cambridge Mass., Harvard UP, 1976-79, Vol. II part I, p. 48; Harley & Woodward, *The History of Cartography*, Vol. II, Book I, pp. 156f., & 220-221; L. Bagrow, *History of Cartography*, London, Watts, 1964, pp. 53-58.

Europe they had mainly from Ptolemy. They knew of the "Franks" vaguely through travellers' reports and diplomacy, and the Islamic urban elite had a stereotyped image of Europeans as brave in battle, but primitive, unhygienic, deceitful, and with very loose morals (especially the women). This product of a cultural Othering process by Muslims, born of ignorance, was largely confirmed as a result of contact with large numbers of "Franks" during the Crusades: Europeans were brave, but primitive and uncivilised, and until they started attacking Islam, not even very interesting.[11] Only later on did the Islamic interest in Europe increase considerably, especially under the Ottomans (Figure 1.6), who regularly laid siege to large parts of it.

Figure 1.7. Liang Chou, World Map, 1593

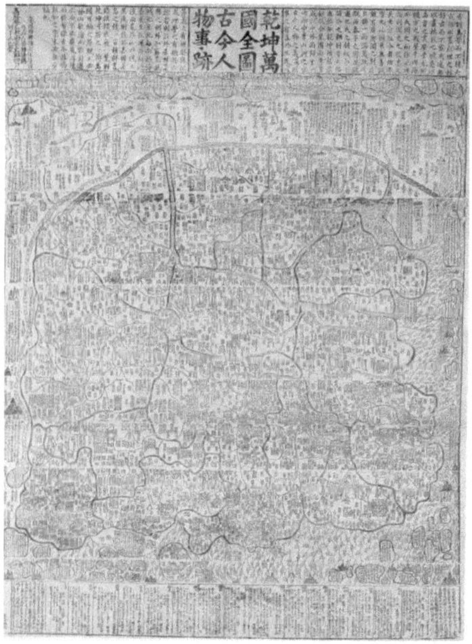

Source: Goss, *The Mapmaker's World*, p. 175.

If Europe's near neighbours in the Levant and the Gulf area were not very preoccupied with the portrayal of Europe on their maps, then it would be unreasonable to expect much more from cartographers further afield, in East and South-East Asia. Before the arrival of the Jesuit Matteo Ricci (1552-1610) in China, the masters of the long tradition of Chinese cartography had little to say about Europe; thereafter, there was

11 Hillenbrand, *The Crusades*, pp. 268-274.

increasing interest, together with the adoption of European scientific and cartographic techniques, especially those derived from Ptolemy. In fact the vast majority of Chinese cartography continued to ignore Europe: it was concerned with mapping China, and went on doing so in a mainly traditional way until the beginning of the twentieth century. Figure 1.7 shows Liang Chou's "A comprehensive map delineating heaven and earth and the myriad countries of the world and modern human affairs", printed in Nanjing in 1593-4. It was the first Chinese world map to incorporate information brought in by Ricci and the Jesuits, but it is not easy to pick out with Western eyes, and Europe is hardly visible. The map remains entirely Sinocentric, continuing to present heaven and earth as entirely dominated by China, and with only the simplest indication of the world beyond its borders.[12]

Figure 1.8. Francesco Sanbiasi SJ, World Map, 1668

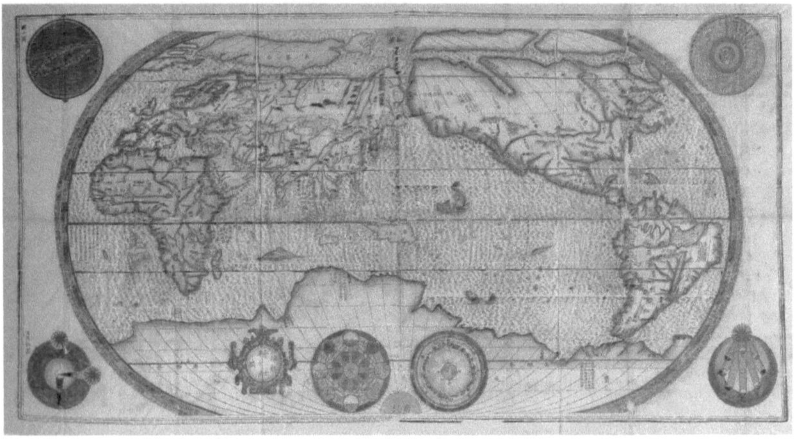

Source: Goss, *The Mapmaker's World*, p. 174-175.

Nevertheless, Ricci's work in spreading Western cartographic techniques impacted by introducing at least to some Chinese observers the notions that the world was spherical, that China was one country amongst very many across the globe, and that it might not after all be at the centre of the world. It also meant that Europe featured more prominently on some East Asian maps of the world, for Ricci's maps were taken from models like those of Ortelius, who was of course specialised in European geography. Figure 1.8 shows a woodblock-printed world map by Francesco Sanbiasi SJ, produced in Guangzhou in about 1648. It

[12] J. Goss, *The Mapmaker's Art: a History of Cartography*, London, Studio Editions, 1993, pp. 174-175.

is based on a much larger map by Matteo Ricci, dating from 1602.[13] These Westernised Chinese world maps were split in the Atlantic, which placed China near the centre, and Europe on the periphery.[14] There was, however, more Chinese interest in Europe than was in evidence in most of the Islamic maps described above. There is in the Museum of Applied Art (Österreichisches Museum für angewandte Kunst, or MAK) in Vienna a beautiful silver and enamel globe from China, of the Qing period, dating from the first half of the eighteenth century.[15] It was purchased by an Austrian diplomat in 1902 in Beijing, possibly from the Imperial Palace. On its surface, the image of Europe is quite accurate and clear, and is considerably larger than China.

Figure 1.9. *Chonha-do* (Under Heaven Map), eighteenth century

Source: National Museum, Seoul, author's photograph.

Chinese cartography was fundamentally influential in Korean map-making, and indeed the standard Korean map of the world, the *Chonha-do* ("under heaven map"), which was printed from woodblocks and bound into atlases throughout the Yi dynasty right up to the twentieth century, was originally based on Chinese texts of the third century AD (Figure 1.9). These traditional maps are Sinocentric, with a massive

[13] *Ibid.*

[14] Harley & Woodward, *The History of Cartography*, Vol. II, Book II, pp. 170-173 & plate 15; Bagrow, *History of Cartography*, pp. 197-203.

[15] MAK Vienna, ref. G01827/1939.

China dominating the world, north at the top, and a small but distended Korean peninsula to the top right. To the left, or west, of the bulk of China are vague shapes which change over the centuries, but which clearly represent Arabia, Africa, and in the extreme top left, Europe. On some of them the Black and Mediterranean Seas are clearly visible, but Europe is so squashed up into the top left-hand corner of the map that it recalls the Arno Peters projection of the 1970s, which makes Europe look (it is said) like a wrinkle at the extremity, or a tiny T-shirt hung out to dry in some far corner of the global garden.[16]

Figure 1.10. *Bankoku-Sozu* Map of the World, 1645

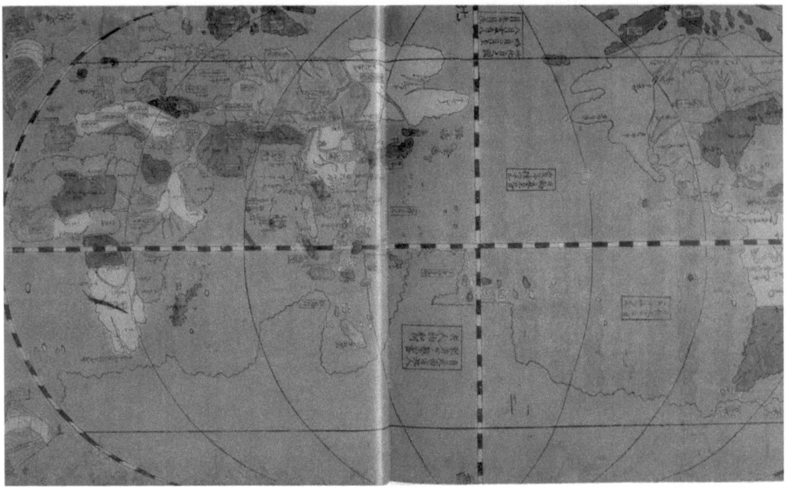

Source: Whitfield, *The Image of the World*, pp. 90-91.

Japanese cartography too was based on Chinese scholarship, and was also influenced by European techniques through the Jesuits and their dissemination of Ricci's work. Figure 1.10 is an example from 1645 of these Westernised *Bankoku-sozu* ("complete maps of the peoples of the world"), which were printed in Nagasaki and spread around the Japanese market in the sixteenth and seventeenth centuries. Dutch maps also heavily influenced the Japanese *Nanban* ("southern barbarian") map genre.

16 Harley & Woodward, *The History of Cartography*, Vol. II, Book II, pp. 246-265 & plate 16; S. McCune, "Korean Maps of the Yi Dynasty", *Korean Culture*, 4/3 (Sept. 1983), pp. 21-31.

Figure 1.11. Japanese Map of the World, Westernised, date unknown

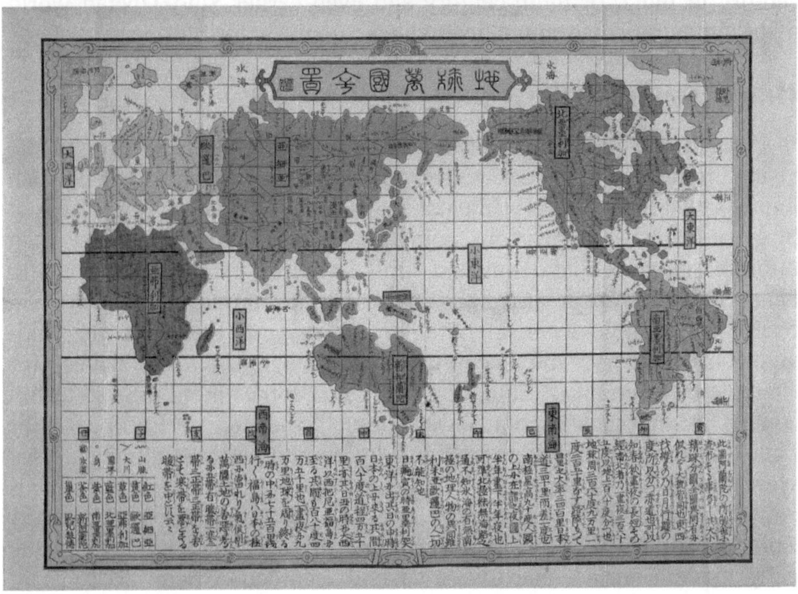

Source: P. van Roojen, *The Agile Rabbit Book of Historical and Curious Maps*, Amsterdam, Pepin Press, 2005, p. 34.

Figure 1.11 shows another example of a Western-influenced Japanese map, of uncertain age; however, the featuring of Australia suggests a date in the nineteenth century. The projection is along the lines of Mercator (Greenland is enormous), but the map is split in the Atlantic, not the Pacific, placing Japan just above the centre in the middle of the world. Europe is reasonably accurate, but squeezed off into a corner, very far from the centre of attention. On the other hand, many other maps, such as Indo-centric, Buddhist world maps, showed Europe as little more than a collection of tiny islands in the top left-hand corner. From the mid-eighteenth century the Dutch influence became even stronger through the promotion of Dutch Studies (*Rangaku*) in Japan, at a time when the Western presence was confined to the Dutch merchants in Nagaski harbour. So despite its isolation from the West, Japan actually had a plenteous supply of European influence on its cartography over several centuries, as well as its own Chinese-derived and religious mapping traditions. The former gave considerable attention to Europe;

the latter hardly at all.[17] From other areas in South and East Asia, traditional religious maps of the world like the Jain charts of the fifteenth century in India, or ninth-century and even earlier Sino-Tibetan world maps, are purely diagrammatic, and have no place for Europe at all.[18]

Other Visual Media

Apart from maps, it is not easy to gain access to the sources of the visual impression of Europe held by the rest of the world, if indeed such impressions existed. Many mental maps and pictures are characterised by great blanks and omissions, sometimes with a purpose but most often through ignorance and disinterest. Only anecdotal evidence has come to hand. Extra-European travellers' tales about Europe exist in written form, for example from an Ottoman embassy to France in the mid-eighteenth century, and accounts of Europe by African slaves, for example by a certain O. Equinnoh of Nigeria at the end of the eighteenth century. Many such accounts were actually penned by Europeans for didactic purposes, and therefore have little to do with a genuine external view: the classic example is Montesquieu's *Lettres Persanes*, which was entirely fabricated.[19] Apparently it is the case that, alongside many European (and Eurocentric) impressions of sub-Saharan Africa in the early modern period, there were several recorded African accounts of the difference between themselves and Europeans, which "grew increasingly sophisticated in their appreciation and estimation of Europe". In recording these differences, and "in discovering Europe, Africans were discovering and redefining themselves", which is of course precisely what the Europeans were doing with their own observations of difference seen from their side of the fence, or the continental divide.[20] There are letters by Indian sepoys which record their impressions of Europe (amongst other things) under the British regime at the time of World War I; Europe figured large in their imagination, but their actual contact with Europeans was limited to the handful of British officers they saw on a daily basis.[21] In similar vein, some work has been done on the

---

17 Harley & Woodward, *The History of Cartography*, Vol. II, Book II, pp. 377-439, especially Figures 11.39 & 11.59 and plate 29; Whitfield, *The Image of the World*, pp. 90-91; Bagrow, *History of Cartography*, pp. 204-206.

18 Whitfield, *The Image of the World*, pp. 30-31; Harley & Woodward, *The History of Cartography*, Vol. II, Book II, pp. 640-643.

19 See M. Harbsmeier, "Character, Identity, and the Construction of Europe", *Ethnologia Europaea*, 29/2 (1999), pp. 5-12.

20 R. Guerrina, *Europe: Histories, Ideas and Ideologies*, London, Arnold, 2002, pp. ix-xiv.

21 D.E. Omissi, "Europe through Indian Eyes", *English Historical Review*, 122/2 (2007), pp. 371-96.

notion of Europe held by Turkish workers in Germany in the 1990s. There was apparently a tension between the reality of daily life in Europe, and the ideal held by these long-term visitors; however, this kind of study is as yet in its infancy, and there is no reference to the visual aspects of their impressions.[22]

Some eighteenth-century carvings from Ecuador of the continents personified would seem to be perfect material for informing us of South American visual impressions of Europe, but they are less than meets the eye. A set of painted carved wooden statues exists in the Museo J. Jijón y Caamaño in Quito,[23] showing the four continents personified as young girls. They are, however, copied from the European engravings of the four continents of around 1,600 which were common in the decorative work of artists especially in the Low Countries, such as Dirck Barendsz, Marcus Gheeraerts, Maarten de Vos and Filips Galle. Europe is crowned, bejewelled and finely dressed, carrying a book; Africa is black with a cornucopia; Asia has a turban and rich clothes, while America is white, carries a bow and arrow, wears a feather headdress and considerably less clothing than the others. This is a rare set of pieces, but it might just as well have been carved in Antwerp or Madrid.

There have been studies of "colonial" and "postcolonial" literatures which deal with the views of postcolonial writers on European civilisation. Onc particular example is an earlier study by Peter Beardsell (the author of the chapter on Latin America in this collection) of Latin American literature and Europe, which also examines some visual material. It includes a section on two paintings from Mexico and Cuba at the time of the Second World War, which engage in an "appropriation of the European gaze".[24] One of them, *Sueño de La Malinche* (1939) by Antonio Ruiz, takes traditional European representations of their own cultural superiority and turns them in on themselves. Beardsell is not dealing with a straightforward rejection of the European influence in postcolonial societies, but more with an "appropriation" of the authority and superiority originally expressed by Europeans in their gaze upon their colonies; he demonstrates the subtlety of the interweaving of the European, the indigenous Indian, the African and other labour imports, and the current Spanish-descended claims on identity in these complex,

22 M. McGowan, "Some Turkish-German Views and Visions of 'Europe'", in B. Axford *et al.* (eds.), *Unity and Diversity in the New Europe*, Oxford, Peter Lang, 2000, pp. 339-369.

23 Inv. nr. 3M4 (39-42); illustrated in L. Schoonbaert *et al.* (eds.), *America Bride of the Sun: 500 Years Latin America and the Low Countries*, Ghent, Imschoot, 1992, pp. 444-445.

24 P. Beardsell, *Europe and Latin America: Returning the Gaze*, Manchester, Manchester UP, 2000, p. 188f.

postcolonial societies. This is a careful and sensitive analysis, and it is extended in his chapter here.

Figure 1.12. Miguel d'Arenzo, "The Rape of America", c. 1990

Source: N. Mirzoeff (ed.), *The Visual Culture Reader*, London, Routledge, 1998, p. 387.

There are further examples to be found. In similar vein, a painting by the Argentinian Miguel d'Arenzo, called "The Rape of America" (c. 1990, Figure 1.12), seeks to turn the tables on traditional European assumptions of superiority. The subject is the traditional one of the myth of the Abduction of Europa, but the power relations are inverted and confused. The Spanish colonial presence in America is represented as Zeus in his bull-form: he is powerful, but intoxicatedly led on by the temptress America, who is confusing the bull with her wiles. There is the atmosphere of a bullfight (a picador lingers to the rear): it is the European presence which is going to be emasculated and ritually humiliated, and his evident power is more tragic than dangerous. This returning of "the gaze" is interesting, for seeing is in this sense a way of negotiating relations between self and surroundings, and the act of perception helps define identity, which is always a complex, hybrid construct, especially in postcolonial situations.[25]

Finally in this admittedly anecdotal survey of non-European visual representations of Europe, we turn to the United States, which has been prolific in its production of images of Europe (many of them enumerated by Ruud Janssens in his chapter in this collection). Without any claim to systematic treatment, some observations on the relationship between the United States and Europe will form a fitting conclusion to this introductory chapter.

[25] J. Thomas (ed.), *Reading Images*, London, Palgrave, 2000, p. 2.

Figure 1.13. Europe and America reversed, frontispiece of the *Massachusetts Magazine* (1790)

Given that the standard image of America – North and South – generated in Europe from the sixteenth to the early twentieth century was of a primitive savage, one might expect some resistance to that notion in American sources, especially from the ex-European but increasingly independent-minded settlements of the east coast of North America. During the period of the American and French Revolutions, American magazines were quite willing to indicate how they could teach "old Europe" a lesson or two about civilisation, especially when it came to political freedoms. The frontispiece of the *Massachusetts Magazine* of 1790 shows a personification of America instructing Europe in basic liberty (Figure 1.13). The two young ladies stand next to a circular metal shield inscribed with the names of all four continents, which has America's name at the top to show that its fortunes are in the ascendant; the turn of the wheel ensures that the time of Asia and Africa will come, while the implication is that Europe is on the way down. Only two of the continents are shown in personified form, and their usual roles are reversed. First, most noticeably and almost uniquely, Europe has no clothes (apart from a modesty drape), while America is fully robed. Both have haloes of enlightenment, but Europe has no crown, nor any other insignia of superiority. She is on the receiving end of instruction in liberty from America, who holds aloft a staff with a Phrygian bonnet of freedom, and points to an open copy of Tom Paine's *The Rights of Man* on the ground, lying next to burst shackles and chains, a mask, and other symbols of freedom and oppression. In the distance, two shadowy figures, one in armour and one in rags, are slipping away; they represent the defeated forces of war and poverty. It is a direct reversal of the usual power relationship portrayed in European iconography between the old and the new continents.[26]

The frontispiece of the *Massachusetts Magazine* was an example of righteous defiance born of successful revolution, and such sentiments continued in certain quarters. However, there was often an ambiguity in American writings about the relative positions of Europe and America, especially in terms of civilisation. America was clearly young, vigorous, fresh, and right in most things (as far as the American pundits were concerned); there was, however, some doubt about Europe. The novels of Henry James are a case in point. An author who was born, raised and educated in the United States but who spent most of his life in Europe, much study has been devoted to his developing feelings about Europe and America.[27] In *The Portrait of a Lady* (1881), which at one level is a story of expatriate Americans trying to come to terms with European civilisation, there is a clear deference to the sophistication of London, Florence, and especially Rome, and some feeling of undefined inadequacy on the part of Americans in the face of all this ancient civilisation. The character of Madame Merle in particular articulates these sentiments.[28]

---

26 Illustration and commentary at J.P. Greene, *The Intellectual Construction of America: Exceptionalism and Identity from 1492 to 1800*, Chapel Hill, University of North Carolina Press, 1993, pp. 132-133. Picture courtesy of Library of Congress, neg. no. LCUSZ62-45522.

27 One example from many: C. Wegelin, *The Image of Europe in Henry James*, Dallas, Southern Methodist UP, 1958. See also the treatment of Henry James in the chapter here by Ruud Janssens.

28 H. James, *The Portrait of a Lady*, Oxford, Oxford UP, 1995 [1881], e.g. pp. 215-218.

Figure 1.14. "New Weight in the Balance",
***Minneapolis Journal*, 1898**

Source: http://www.csubak.edu/~gsantos/cat20.html, consulted 7 June 2005.

If we turn to cartoons, and in particular to one from the *Minneapolis Journal* shown in Figure 1.14, then the Spanish-American War of 1898 appears to have been the definitive occasion on which a mature United States established its claims on the world stage as a fully fledged imperial power, at the direct expense of one of the old European imperial nations, Spain. American victories resulting in the loss of the remnants of the Spanish empire – Puerto Rico, Cuba, and the Philippines – occasioned a wave of patriotic cartoons which put Uncle Sam and Teddy Roosevelt as personifications of rampant America on a par with the personifications of the European powers (John Bull, the Kaiser, the Tsar, the Austrian Emperor), and were often anything but complimentary about Europe.

Figure 1.15. "Spaniards Search Women on American Steamers", press drawing by Frederick Remington, c. 1900

Source: http://www.humboldt.edu/~jcb10/spanwar.shtml, consulted 7 June 2005.

Spain was especially singled out for detrimental portrayal, as a tired, lecherous old bandit: one cartoon in particular, shown in Figure 1.15, showed a naked young (American) woman being man-handled by Latin types on a boat deck. Uncle Sam was compared favourably with Cecil Rhodes, as America emulated European imperialism, even to the extent of taking on an Americanised version of the "White Man's Burden" as a "euphemism for imperialism, that seemed to justify the policy as a noble enterprise". This, then, was an assertion of American identity in a period of jingoism, which tended to downgrade the European Other, but was blend of imitation of Europe's imperialism and disdain for its exhausted venality, especially in the case of Spain.[29]

[29] See a collection of various cartoons available on the World Wide Web, at the following sites: http://www.temple.edu/history/trmonroecart.jpg, http://www.csubak.edu/~gsantos/cat20.html; http://www.oah.org/pubs/magazine/1898/martinez-lesson-pdf; http://www.indiana.edu/~jah/teaching/archive/2002_03/sources.shtml; http://www.nypl.org/research/chss/epo/spanexhib/; http://www.smplanet.com/imperialism/cartoon.html; http://www.humboldt.edu/~jcb10/spanwar.shtml, all consulted 12 April 2003; and http://www.boondocksnet.com/gallery/cartoons/1898/, consulted 12 March 2003. I am grateful to Yu-Chun Kao, an MA student at the University of Amsterdam in 2002/3, for bringing these cartoons to my attention.

Figure 1.16. Daniel Chester French, the Continents: Europe, America, Africa, New York, 1907

Author's photograph.

At the same time, however, in less jingoistic mood, the United States appeared happy to share the standard, more Eurocentric view of the continental hierarchy at the beginning of the twentieth century, with the proviso of course that the Europe-derived noble qualities of America be given special emphasis, alongside her own more youthful and vigorous characteristics. Outside the old Customs House in New York's Bowling Green (ironically, perhaps, now the New York part of the National Museum of the American Indian), there stand four majestic statues of the continents (Figure 1.16), sculpted by Daniel Chester French (who also did the Lincoln Memorial in Washington). Europe is a woman in her prime, enthroned on the right of the main portal of this grand neo-classical building, leaning back comfortably and at ease with her position in the world. Her pose is not unlike her sculptor's more famous characterisation of Abraham Lincoln, assured in his massive armchair in his massive monument in Washington DC. At the New York Customs House, Europe is crowned and has a huge globe to show her pervasive influence in the high noon of New Imperialism; ship's chains and anchors show her command of the oceans, while a charger carved on the throne and a hooded figure with a book and globe refer to her martial and intellectual talents. Meanwhile, white America sits on the other side of the entrance, without crown or throne. She is younger than Europe, and leans forward in eager anticipation of life's challenges, while a young naked male kneels at her side, probably a personification of native America, but also linked to Hermes and trade (the other continents are banished to the outer reaches of the building; on the far left, we can just see Africa as a non-negro, northern African woman, with two bound slaves next to her in anguish and agony. This is to a large extent a Eurocentric image on the eve of the First World War, despite the fact that it was commissioned, sculpted and then sited in New York's urban public space. Europe remains regal, and the senior figure. This is not an isolated image of the time in America. For example, two statues flanking the main door of the Surrogate's Court building, near City Hall in New York, show America and Europe. The statues date from the early twentieth century, and America is shown with Indians and settlers. On the other side of the gate is Europe, also resembling Britannia: she holds a globe, and is accompanied by an armed soldier whose shield reads, "Honi soi qui mal y pense", the motto of the British Grenadier Guards, which also appears on the insignia of the Order of the Garter, founded in the fourteenth century by Edward III. Again, this is a basically Eurocentric image, associated with the English legal order. If more evidence of the point were needed, the splendid staircase in the vast, ornate hall of the Library of Congress building in Washington DC, of around 1900, has representatives of the four continents personified as children playing with globes on the balustrade: all four are portrayed as *putti*, but America, in its seat of learning, is still shown as a babe with an Indian feather bonnet, accompanied by an alligator.

Even in the 1970s and beyond, the USA continued to grapple with the image of Europe, *vis-à-vis* its own identity. A book of 1973 entitled *America and the Image of Europe*[30] was still measuring America against the old spectre, rather than the other way around. It and others like it show the perennial importance of the Other in the formation of identity, including national identity, and it is interesting that it talks in terms of a collective "European" identity as a contributor to American ideas about their own essential qualities. And as with all the American images, it shows that, despite the increasing supremacy of American power in field after field, especially in the twentieth century, Europe still counted for a great deal.

[30] D.J. Boorstin, *America and the Image of Europe: Reflections on American Thought*, Gloucester Mass., Smith, 1976.

Figure 1.17. Dave Brown, "Bush's New World Order Map",
***The Independent* (8 April 2003)**

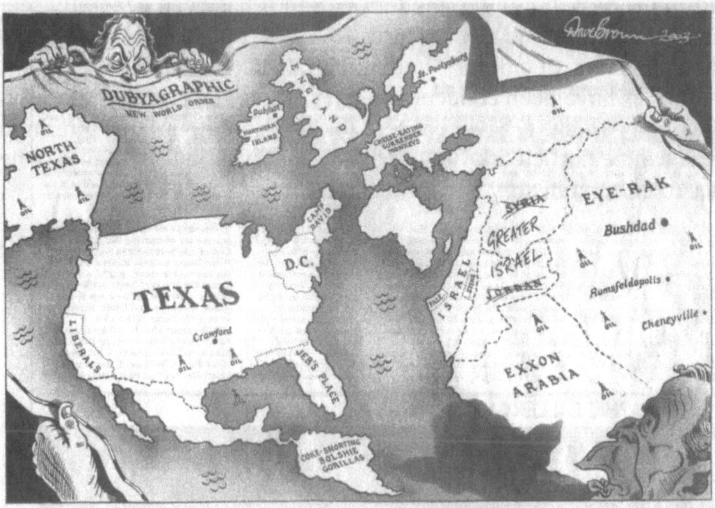

Now, of course, in the twenty-first century, the American view of Europe has moved on: George Bush saw it as the home of the "cheese-eating surrender monkeys", ever since some nations were less than happy about his policy in Iraq. The cartoon in Figure 1.17 is actually a European one, by the inimitable Dave Brown in *The Independent*, but the message is clear: the United States is certainly not in awe of Europe any more, far from it.

Conclusion

Before about 1900, the picture we have gleaned of others' views of Europe is very fragmentary, but we can note that, unless there was a specific and favoured interaction with Europe, as there was in the Japanese case, many parts of the world largely ignored it. Where there was a colonial or postcolonial context, then feelings were always mixed: rejection and reversal of dominance were almost always tempered with imitation, rivalry, or recognition that the European connection had affected and modified (post-) colonial feelings. The concept of "hybridity" may well be a useful tool here. Crafting the image of Europe has often been a complex process, from both within and without.

These fleeting visual images shown here have, it is to be hoped, provided us with some themes to take with us in looking at other impressions of Europe, especially in the twentieth century. Firstly, when it comes to identity formation, representations of Others are just as important as one's images and perceptions of oneself. Secondly, Europe was ignored as simply unimportant and uninteresting by much of the world in the past, but in the modern period it has been generally seen as a force to be reckoned with. Third, from the outside, the perception of Europe in terms of military and imperialistic power has been just as important as have been concepts of civilisation. And finally, these visual images, on which I have chosen to concentrate in this preliminary survey, can be particularly illuminating, and even indispensable, when it comes to cultural history and mutual perceptions.

PART II

FROM TOTALITARIAN EUROPE

CHAPTER 2

Heroes and Merchants

Joseph Stalin and the Nations of Europe[1]

Erik VAN REE

Introduction

In his *Händler und Helden* (1915), the German economist and sociologist Werner Sombart interpreted the Great War as an existential battle not just between nations, but between cultures and world views. According to Sombart, West European civilisation was based on the ideas of 1789 and on commercial values, which he identified with the Jewish spirit. The typical West European was a merchant, exclusively interested in what life could offer him in terms of goods and comfort. In contrast, Germany was a nation of heroes, who were prepared to sacrifice themselves for higher ideals.[2]

Sombart's book contributed to the radical right-wing tendency of the so-called "Conservative Revolution". The Conservative Revolutionaries started out as a Romantic, *fin de siècle* phenomenon. Leading lights such as Oswald Spengler, Arthur Moeller van den Bruck and Ernst Jünger deplored the decline of traditional society based on hierarchy and authoritarianism, and were fundamentally opposed to capitalist commer-

1 This is a revised version of an article published in *Kritika: Explorations in Russian and Eurasian History*, 8/1 (2007). I am grateful to the editors for permission to publish the revised version here.

2 W. Sombart, *Händler und Helden. Patriotische Besinnungen*, Munich/Leipzig, Duncker & Humblot, 1915; J. Herf, *Reactionary Modernism. Technology, Culture, and Politics in Weimar and the Third Reich*, Cambridge, Cambridge UP, 1984, p. 144. For an illuminating analysis of Sombart's ideas on heroes and merchants: R.P. Sieferle, *Die Konservative Revolution. Fünf biografische Skizzen*, Frankfurt M., Fischer Taschenbuch, 1995, pp. 83-92. The hero-merchant duality comes from Arthur Moeller van den Bruck: O. Schüddekopf, *Linke Leute von rechts. Die nationalrevolutionäre Minderheiten und der Kommunismus in der Weimarer Republik*, Stuttgart, Kohlhammer, 1960, p. 36. Moeller van den Bruck contrasted the "young peoples" (Germany and Russia) with the "old", materialist West: Herf, *Reactionary Modernism*, p. 37; S. Breuer, *Anatomie der konservativen Revolution*, Darmstadt, Wissenschaftliche Buchgesellschaft, 1993, p. 40.

cialism and the liberal and egalitarian principles of the French Revolution. Their goal was a new society ruled by a heroic warrior elite. The lost war deepened their apocalyptic rage.[3]

Surprisingly, though he matured and operated within a very different political tradition, Joseph Stalin's views of the mentalities of the nations of Europe were very similar to Sombart's.[4] This is not to say that the Soviet dictator was of one mind with the Conservative Revolutionaries in all respects. He would never have classified "the ideas of 1789" as decadent. For him, they were precisely the higher values for which one should sacrifice oneself. It was the avaricious capitalists with their commercial mentality who had betrayed the revolutionary ideals. The Communists had inherited the mantle of the French Revolution.

In his own way, Stalin might even be called "pro-Western", in the sense that he was an ardent moderniser and a jealous admirer of Western progress. Typically, in February 1947 he remarked to Sergei Eisenstein that it was impossible to deny the progressive role of the Christianisation of Russia. It "marked the Russian state's shift towards joining up with the West, instead of an orientation towards the East".[5] Until his death it remained Stalin's fond goal to overtake the West in terms of civilisation and technology. Russia's backwardness deeply troubled him, as appears from his well known speech of February 1931 about the causes of Russia's historical defeats.[6] The Soviet leader betrayed his insecurity by remarks like that made to a Polish delegation in April 1945: "the Polish workers are good workers. They are more cultured than ours. The proximity of the West makes itself felt".[7]

Stalin did his best to convince himself that the East Europeans were culturally no less developed than their Western counterparts. At a

---

3 See on the Conservative Revolution: A. Mohler, *Die Konservative Revolution in Deutschland. 1918-1932*, Stuttgart, Friedrich Vorwerk, 1950; Schüddekopf, *Linke Leute*; L. Dupeux, *"Nationalbolschewismus" in Deutschland. 1919-1933: Kommunistische Strategie und konservative Dynamik*, Munich, Beck, 1976; Herf, *Reactionary Modernism*; Breuer, *Anatomie*; Sieferle, *Die Konservative Revolution*; R. Woods, *The Conservative Revolution in the Weimar Republic*, Basingstoke, Palgrave, 2001.

4 For an analysis of the similarities between right and left-wing anti-Westernism, see: I. Buruma & A. Margalit, *Occidentalism. The West in the Eyes of its Enemies*, New York, Penguin, 2004.

5 A. Artizov & O. Naumov (eds.), *Vlast' i Khudozhestvennaia Intelligentsiia. Dokumenty TsK RKP(b)-VKP(b), VChK-OGPU-NKVD o Kul'turnoi Politike. 1917-1953 gg.*, Moscow, Demokratiia, 1999, p. 613.

6 I.V. Stalin, *Sochineniia*, Vol. 13, Moscow, Gosudarstvennoe izdatel'stvo politicheskoi literatury, 1951, p. 38.

7 V.A. Malyshev, "Dnevnik narkoma. 'Proidet desiatok let, i eti vstrechi ne vosstanovish' uzhe v pamiati'", *Istochnik*, 5 (1997), p. 129.

meeting of economic leaders and Stakhanovists in the Kremlin in October 1937, he admitted that the Russians remained "culturally behind", but in terms of *political* culture they were ahead. "In the West people don't throw cigarette-stubs on the floor, but the working people over there are slaves of capital".[8] He told Andrei Gromyko that "the Bulgarian people are not at all at a lower level of general development than the Germans. In times long ago, when the ancestors of the Germans still lived in the woods, the Bulgarians already had a high culture".[9]

But he didn't really believe it himself. Stalin operated with a scheme in which Asia stood for barbarism, violence and rudeness, and Europe for refined culture, with the Russians somewhere in the middle on the ladder of civilisation, below the Europeans. His dim view of Asians compared to Europeans showed when he described himself. In December 1937 Stalin met with some of his comrades in People's Commissar Klim Voroshilov's Kremlin apartment. He said, "Excuse me, comr. Dimitrov, for interrupting you, I am no European but a Russified Georgian-Asian".[10] In August 1939 he told German Foreign Minister Joachim von Ribbentrop that force was the only language the Japanese – "these Asians" – understood. "Moreover, I am one myself and I know what I'm talking about".[11] During the hysterical anti-cosmopolitan campaigns of his last years, the dictator insisted that the Russians had always been the world's greatest pioneers. But the very fact that he reiterated so emphatically Russian "priority" confirmed that he continued to realise that this pre-eminence had not yet been achieved.

This being said, the parallels between Stalin's views and Sombart's were indeed striking. Like the latter, the Soviet leader celebrated the heroic spirit of sacrifice, and he acknowledged that Western culture overvalued comfort compared to struggle. Just like Sombart he abhorred the commercial spirit. For a Communist, Stalin put a remarkable emphasis on the cultural, as opposed to the socio-economic, downside of capitalism. He criticised Europe as a capitalist *culture*, condemning it as a *morally* defective system. Capitalism denied people the possibility of becoming heroes and thereby sapped its own vital strength.

On 17 March 1938 Stalin received the polar explorer I.D. Papanin and three of his comrades in the Great Kremlin Palace. They had been

---

8 B.A. Nevezhin, *Zastol'nye Rechi Stalina. Dokumenty i Materialy*, Moscow, AIRO-XX, 2003, p. 128.

9 A.A. Gromyko, *Pamiatnoe*, Vol. 1, Moscow, Izdatel'stvo politicheskoi literatury, 1988, p. 201.

10 Nevezhin, *Zastol'nye Rechi*, p. 151.

11 G. Hilger, *Wir und der Kreml. Deutsch-sowjetische Beziehungen 1918-1941. Erinnerungen eines deutschen Diplomaten*, Frankfurt, Alfred Metzner, 1955, p. 291.

saved from a floe of ice floating in the open sea and had been safely delivered to port by the ice-breaker *Ermak*. Stalin was in high spirits and remarked,

> There in the West, in France, in Germany, in England, in America they make no heroes [...] The Norwegians turned to us with a proposal of assistance. [...] we know that these bastards [...] cannot offer any significant assistance. They only pretend. For what advantages, speaking in the terms of the foreigners [*po-inostrannomu*], what earnings would be in it for them? [...] we decided to spare no amount of money and no icebreakers. [...] what is an act of heroism, what is it worth? Not a single American, not a single Frenchman, not a single Englishman will be able to tell you, because he has only one valuation: the dollar, the sterling, the franc.[12]

Here we have Sombart's contrast of the easy-going, mercantile societies of the West, with their taste for profit, versus the (Soviet-Russian not German) heroic society driven by the spirit of sacrifice. What Stalin said in March 1938 was not exceptional. Time and again his contempt for the mercantile spirit showed, as for example during the war when Yugoslav Communist Milovan Djilas offered to repay him for his arms deliveries. The Soviet leader became unexpectedly angry: "You insult me. You are shedding your blood, and you expect me to charge you for the weapons. I am not a merchant, we are not merchants".[13] And although he acknowledged their backwardness, Stalin admitted to an intense admiration for the heroic qualities of the Russian people. Klaus Mehnert, Terry Martin and David Brandenberger, among others, have discussed his conception of Russian heroism.[14] The Soviet leader saw the Russians as a nation by character opposed to slavishness and traditional routines, active, steadfast, revolutionary and with a talent for waging war.[15]

---

12 Nevezhin, *Zastol'nye Rechi*, pp. 189-191.

13 M. Djilas, *Conversations with Stalin*, New York, Harcourt, Brace & World, 1962, pp. 63-64.

14 K. Mehnert, *Stalin versus Marx. The Stalinist Historical Doctrine*, London, George Allen & Unwin, 1952; T. Martin, *The Affirmative Action Empire. Nations and Nationalities in the Soviet Union, 1923-1939*, Ithaca, Cornell UP, pp. 451ff.; D. Brandenberger, *National Bolshevism. Stalinist Mass Culture and the Formation of Modern Russian National Identity, 1931-1956*, Cambridge Mass., Harvard UP, 2002.

15 See *On the Foundations of Leninism* (1924), in *Sochineniia*, Vol. 6, p. 186. In 1930 Stalin criticised the proletarian writer Dem'ian Bednyi for suggesting that the Russians were still the same lazy Oblomovists of old (*Sochinenia*, Vol. 13, p. 25). For similar statements: Artizov & Naumov, *Vlast'*, p. 136; *Sochineniia*, Vol. 13, pp. 110-111; Nevezhin, *Zastol'nye Rechi*, pp. 44-45. For his famous May 1945 toast to the Russian nation, see *Sochineniia*, Vol. 2 [XV], pp. 203-204. When Charles de Gaulle asked Stalin why he always worked so hard, he answered that this was "a bad Rus-

This chapter takes a closer look at Stalin's assessment of the national characters of various European nations. The materials brought together here come from various times and places, but that does not render their presentation arbitrary. On the contrary, it is remarkable to see how consistent Stalin was in attributing character traits to these nations. To be sure, when he made his high or low opinions of nations known to others, Stalin often seemed to operate with a hidden political agenda. Nevertheless, his assessments hardly varied over time but generally remained the same, independent of the occasion on which he chose to present them.

The Bolshevik leader's remarks were for the most part in the nature of loose talk. He felt uncomfortable seeing his views of specific nations laid down in official documents, especially if they were depreciatory. He ventilated such views preferably in casual conversations with party comrades; with state officials or delegations from segments of Soviet society; with Soviet and foreign journalists, writers and other artists; and with visiting foreign Communists, diplomats and statesmen. But despite the "table talk" character of Stalin's remarks, the pattern in them is clear enough.

I shall conclude that the March 1938 speech about the heroic Soviet people and decadent Westerners did, indeed, represent Stalin's thinking all along. Like the Conservative Revolutionaries, he routinely operated with powerful national stereotypes, dividing up the European nations according to the archetypal images of the hero and the merchant. He saw the continent divided up into a heroic but less developed East and a developed but commercialist and cowardly West. I will argue that, despite his strong views about national characters, Stalin's essentialism did not originate from true racism. For him, national characters remained rooted in circumstances rather than in genetics, and they changed over time. His low view of the flabby Western character was rooted in his anti-capitalist ideology and in his lurking suspicions of the dangers and temptations inherent in socio-economic success. Finally, his praising and slighting of European nations should not principally be understood as the fruit of an intellectual interest, but as the product of strategic considerations and of his understanding of politics as mobilisation. The Soviet dictator's aim was to instil a heroic fighting mentality into the Soviet nations and their allies.

---

sian habit", "Oshibki imeiutsia i u menia, i u moikh sotrudnikov". "Zapis' besedy I.V. Stalina i Sharlia de Gollia", *Istochnik*, 5, 1996, p. 106. The Soviet leader was exorcising the Oblomovist spirit.

The Merchant Nations

Among the peoples living in Europe about whom Stalin had the strongest views were the Jews. They were for him the ultimate commercial nation. According to the Menshevik R. Arsenidze, he described them in 1905 as follows: "Just try and work with them. They are no good in a battle, nor for having fun together. Cowards and petty dealers!".[16] In his "Marxism and the National Question" (1913), he noted that there were few Jewish peasants. The Jews typically served "alien" nations "as industrialists and traders, as well as people of the liberal professions".[17]

Stalin felt a profound contempt for the supposedly money-loving Jews. In December 1941 he told a Polish delegation that before the First World War he once tried to travel illegally to Poland and arrived in an unknown border town. Some Jews offered to help him. "But I [...] did not trust these Jews. It was written on their faces that they were prepared to hand me over to the Russian gendarmes for money. In the end I found a Pole with an honest face".[18] At Yalta in February 1945 he informed Roosevelt that Jews were in his opinion "middlemen, profiteers and parasites".[19] Though they could be convinced to take up agriculture, they remained "natural traders" with a preference for the cities.[20] At the end of his life, when Soviet anti-Semitism developed into a campaign, the same point was on Stalin's mind. In December 1952 he told the party Presidium: "All Jews are nationalists [...] The Jewish nat[ionalists] believe that the United States saved their nation (out there you can become rich, bourgeois etc.)".[21]

Stalin's opinion of the Americans in some ways paralleled his views of the Jews. In 1924 he praised them for their "efficiency", though that

---

16 R. Arsenidze, "Iz vospominanii o Staline", *The New Review. Novyi Zhurnal*, 72, June 1963, p. 221.

17 *Sochineniia*, Vol. 2, pp. 333-334.

18 Nevezhin, *Zastol'nye Rechi*, p. 302.

19 C.E. Bohlen, *Witness to History. 1929-1969*, London, Weidenfeld & Nicolson, 1973, p. 203.

20 *Foreign Relations of the United States. Diplomatic Papers. The Conferences at Malta and Yalta. 1945*, Washington, US Government Printing Office, 1955, p. 924. Stalin's closest Jewish comrade Lazar' Kaganovich did not admit that his boss was an anti-Semite, but he did point to the latter's concern that there were too many "petty-bourgeois elements" among the Jews: F. Chuev, *Tak Govoril Kaganovich. Ispoved' Stalinskogo Apostola*, Moscow, Otechestvo, 1992, p. 128. According to Molotov, Stalin appreciated the following qualities in the Jews: "diligence, unanimity [*spaiannost'*], political activism". F. Chuev, *Sto Sorok Besed s Molotovym. Iz Dnevnika F. Chueva*, Moscow, TERRA, 1991, p. 274.

21 Malyshev, "Dnevnik narkoma", pp. 140-141.

easily degenerated into something "narrow and unprincipled".[22] The Americans were hard workers but they were not very far-sighted. Stalin told Emil Ludwig in 1931 that he appreciated their "unassumingness" [*prostota*] and their "democratism", traits to be explained by the absence of a feudal aristocracy in the USA.[23] But at the end of his life Stalin lost much of his respect for the Americans. In August 1951 he told Chinese Premier Zhou Enlai that he did not expect the Korean War to turn into a global conflagration. The Americans preferred to rely on their nuclear weapons and air force instead of on the infantry.

> Americans are merchants. Every American soldier is a speculator, occupied with buying and selling. [The] Germans conquered France in 20 days. It's been already two years, and [the] USA [have] still not subdued little Korea. What kind of strength is that? America's primary weapons [...] are stockings, cigarettes, and other merchandise. They want to subjugate the world, yet they cannot subdue little Korea. [...] They are fighting with little Korea, and already people are weeping in the USA. What will happen if they start a large-scale war? Then, perhaps, everyone will weep.[24]

Like the Jews, the Americans were a nation of traders and cowards – not soldiers.

The theme returned in Stalin's characterisation of the French and the British. There are the intriguing remarks, probably made by Stalin and written down by party secretary Andrei Zhdanov before the concluding of the Molotov-Ribbentrop Pact in 1939: England and "syphilitic Europe" allegedly hoped to send the "tigers", Nazi Germany, to the East.[25] The image of "syphilitic Europe" expressed a perceived weakness and lack of resolve combined with shrewd scheming to let others do the fighting. In August 1939 Stalin told Ribbentrop that British power was based on bluff, stupidly taken for real strength by other nations. The French were also too weak to wage war. Stalin once again vented his conviction that merchants have no backbone, noting as he did "that the Anti-Comintern Pact had in fact frightened principally the City of London and the small British merchants".[26] In his discussion with

---

22 *Sochineniia*, Vol. 6, pp. 187-188.

23 *Sochineniia*, Vol. 13, pp. 114-115.

24 "Stalin's Conversations with Chinese leaders. Talks with Mao Zedong. December 1949-January 1950, and with Zhou Enlai, August-September 1952", *Cold War International History Project Bulletin*, 6-7, 1995/96, pp. 12-13.

25 L.A. Bezymenskii, "Sekretnyi pakt s Gitlerom pisal lichno Stalin", *Novoe vremia*, 1, 1998, p. 31.

26 Hilger, *Wir und der Kreml*, p. 291; R.J. Sontag & J.S. Beddie (eds.), *Nazi-Soviet Relations. 1939-1941. Documents from the Archives of the German Foreign Office as Released by the Department of State*, New York, Didier, 1948, pp. 74-75.

Ribbentrop in September 1939, he described Britain, America and France as "satiated states" and spoke of "typical French claptrap".[27]

Addressing graduates of the Academy of the Red Army on 5 May 1941, Stalin remarked that after the First World War, Britain and France had begun to neglect their armed forces. Their victory had made them overconfident. The French had now been defeated by the Germans because they had lost interest in their own army. The people had started to look down on the commanders "as on lame ducks [*neudachnikov*], the lowest people [*poslednikh liudei*], who were obliged to go into the army because they had no factories, plants, banks, shops".[28] His contempt for the soft French came out once more when Stalin's Hungarian comrade Mátyás Rákosi paid him a compliment. He sharply interrupted him and said, "Listen, I'm no Frenchman, you shouldn't pay me compliments!"[29]

Stalin also despised other West European nations for their lack of character. We had already heard him doubting Norwegian valour in the rescue of the Russian polar explorers. In March 1945 he told Czechoslovak President Eduard Beneš that the Slavs had suffered most in the war. "Did France suffer more? No. The French opened the front for the Germans. [...] And Belgium and Holland immediately put their feet into the air and lay down for the Germans".[30] In October 1945 Stalin had this to say to a delegation of the Finland-USSR society:

> Compare, for instance, Finland and Belgium. The Belgians regard the Finns a semi-peasant people without culture. But the Finnish people develop their country and wouldn't behave the way the Belgians behaved during the war. The Belgians considered themselves one of the most cultured peoples of Europe, but when war came they surrendered.[31]

In the years after the war Stalin became painfully aware of the attractiveness of Western culture. But he stuck to his view that this culture was fatally flawed. Expressing his unease about the fact that millions of Soviet citizens had seen life abroad as soldiers, he told Zhdanov in September 1947, "There is only one thing which these gentlemen who long for the "Western way of life" cannot explain: why

27 I. Fleischhauer, "The Molotov-Ribbentrop Pact: the German Version", *International Affairs*, 8, 1991, pp. 128-129.

28 A.A. Pechenkin, "'Sovremennaia armiia – armiia nastupatel'naia'. Vystupleniia I.V. Stalina na prieme v Kremle pered vypusnikami voennykh akademii. Mai 1941 g.", *Istoricheskii arkhiv*, 2, 1995, p. 28; Malyshev, "Dnevnik narkoma", pp. 115-116.

29 "'Videl, kak voznikaet kul't lichnosti. Mat'iash Rakoshi o Staline i o sebe", *Istochnik*, 1, 1997, p. 119.

30 Malyshev, "Dnevnik narkoma", p. 128.

31 Rossiiskii Gosudarstvennyi Arkhiv Sotsial'no-Politicheskoi Istorii [RGASPI], f.558, op. 1, d. 5379, l. 8.

we beat Hitler".[32] Attractive though it was, the soft West was less equipped for war than the Russians were.

The Hero Nations

Stalin admired the East European nations for their military qualities. We have seen him calling the Finns brave, and he also admired the Yugoslavs for their fighting spirit. As he told his comrade Josip Broz-Tito in May 1946, they were a "strong people" and a "hardy nation".[33] So were the Slovaks. In December 1943 Stalin advised Beneš to keep a firm grip on these combative people in order to prevent their separatism from undermining the unity of the state. He warned that during the fighting in the North Caucasus his military men had noted "that the Slovak soldier is a good [soldier]".[34] The Albanians, too, were a fierce people. In 1949 Stalin told Communist leader Enver Hoxha that the survival of the Albanian language "proves the endurance of your people, the great strength of their resistance to assimilation despite the storms that have swept over them".[35] But he seems to have appreciated the Albanians less than the Slavs.[36]

The oddest hero nations were the Poles and the Germans. In October 1923 the Politburo discussed whether Poland could be successfully influenced by applying pressure on them. Stalin believed they could not. He scribbled the following note on a piece of paper: "You can frighten the Latvians, push them to the wall etc. But you can't do that with the Poles. […] Isolate the Poles. Buy off (and scare) the Latvians. Buy off the Romanians".[37] The Poles were not for sale, and could not easily be scared off either. In December 1941 Stalin told Polish General Anders

32 D.T. Shepilov, "Vospominaniia", *Voprosy istorii*, 5, 1998, pp. 11-12.

33 L. Gibianskii, "The Soviet Bloc and the Initial Stage of the Cold War: Archival Documents on Stalin's Meetings with Communist Leaders of Yugoslavia and Bulgaria, 1946-1848", *Cold War International History Project Bulletin*, 10, 1998, pp. 120-121.

34 V.V. Mar'ina, "Peregovory E. Benesha v Moskve (dekabr' 1943 g.)", *Voprosy Istorii*, 3, 2001, p. 15.

35 E. Hoxha, *With Stalin. Memoirs*, Tirana, 8 Nëntori, 1979, p. 120.

36 In a discussion with Milovan Djilas (*Conversations*, p. 79) about the Albanians in June 1944, he said, "I had hoped that the Albanians were at least a little Slavic". On 3 January 1946 Stalin had a meeting with Leo Krzycki, chairman of the All-Slav Congress of the United States, who informed him that the American Slavs were very fertile and gave the American army many soldiers, as well as donating unusually large amounts of blood to the Red Cross for the war effort. Stalin commented that "the Slavic nations are young and healthy". L. Maksimenkov, "Slavianskie natsii – molodye i zdorovye", *Nezavisimaia Gazeta*, 31 Jan. 2006.

37 D. Volkogonov, *Sem' vozhdei. Galeria liderov SSSR v 2-kh knigakh*, Moscow, Novosti, 1995, p. 234.

that he saw the ethnic Poles as the very best soldiers, better than the White Russian, Ukrainian or Jewish Poles.[38] In April 1944 he told a Polish visitor that "the Polish people are good people. The Poles are brave warriors".[39] In early 1945 Andrei Gromyko heard him say that the Russians were the "most steadfast" soldiers of the world, followed by the Germans and Poles in that order.[40] But though Stalin admired the Poles, he did not like them. They were an unruly people of natural anarchists whose heroism easily degenerated into heroics. In a way they liked fighting *too* much, taking it to the point where they fought among themselves. As Stalin expressed it to Winston Churchill in October 1944, "if a Pole is on his own, then he gets into a quarrel with himself".[41] At the Yalta Conference in February 1945 he shared his thoughts about the Polish character with Churchill once again: "There are some very good people among the Poles. They are good fighters. Of course, they fight among themselves too".[42]

The Germans were the strangest people of all. Stalin told Emil Ludwig in 1931 that he loved no nation as much as them. They gave the world Marx and Engels.[43] Stalin nurtured an immense respect for the Germans as soldiers. In his conversation with British Ambassador Stafford Cripps in July 1940, he called them a "military people".[44] When De Gaulle once asked him what he thought of the Germans, Stalin answered that "there are good people in Germany but not many". But he respected them as warriors: to be killed, they literally had to be cut to pieces.[45] In a conversation with Beneš on 28 March 1945 he said: "*I hate the Germans*. But hatred must not hinder us from objectively appreciating the Germans. The Germans are a great people. Very good technical people [*tekhniki*] and organisers. Good, courageous, born soldiers".[46] They could even overdo it. In December 1948 Stalin accused Socialist Unity Party of Germany (SED) leaders of operating without tactical *finesse*. "The ancient Teutons entered battle with the Romans

38 Nevezhin, *Zastol'nye Rechi*, p. 300. See also T.V. Volokitina *et al.*, *Vostochnaia Evropa v Dokumentakh Rossiiskikh Arkhivov. 1944-1953 gg.*, Vol. 1, *1944-1948 gg.*, Novosibirsk, Sibirskii khronograf, 1997, p. 40.

39 Volokitina *et al.*, *Vostochnaia Evropa*, p. 40.

40 Gromyko, *Pamiatnoe*, pp. 201-202.

41 "Zapis' besedy tov. I.V. Stalina s Cherchillem. 9 oktiabria 1944 g. v 22 chasa", *Istochnik*, 2, 2003, p. 49.

42 *Foreign Relations*, p. 853.

43 *Sochineniia*, Vol. 13, p. 115.

44 V.I. Sipols, "Missiia Krippsa v 1940 g. Beseda so Stalinym", *Novaia i noveishaia istoriia*, 5, 1992, p. 34.

45 "'Oshibki imeiutsia i…", pp. 105-106.

46 Malyshev, "Dnevnik narkoma", p. 128.

naked, but they suffered losses". Stalin acknowledged that the Teutons were "very brave", but not shrewd enough to take on the Romans. The German Communists were, he believed, like them.[47]

In 1913 Stalin classified the Germans with Magyars and Russians as nations "the most suited to organise states".[48] In part he attributed this to their relatively high level of development. At the tenth party congress in 1921 he noted that the multinational empires of Eastern Europe had been formed "with one, more developed nation at the head".[49] In April 1945 he told Milovan Djilas that the Germans "will recover, and very quickly. That is a highly developed industrial country with an extremely skilled and numerous working class and technical intelligentsia".[50]

But German resilience was also a matter of psychology. In January 1943 Stalin told Comintern leader Georgii Dimitrov that the majority of the German workers apparently enjoyed being a "*ruling nation*".[51] In November of that year he informed Churchill and Roosevelt that there was no hope of reforming the German national character.[52] As he told Beneš in December 1943, the Germans were "under the influence of state fetishism". This "strong and talented people" could only imagine their relations with other nations in terms of subjecting them. The Germans would go to war again. "You can't change the Germans" [*Nemtsev ne izmenit'*].[53]

Apart from their great heroic qualities, however, the Germans suffered from a serious character flaw. Oddly, they were natural slaves as much as natural masters. Stalin told Djilas in June 1944 that he remem-

---

47 "'Nuzhno idti k sotsializmuy ne priamo, a zigzagami'. Zapis' besedy I.V. Stalina s rukovoditeliami SEPG. Dekabr' 1948 g.", *Istoricheskii arkhiv*, 5, 2002, p. 8. In a discussion with a Polish government delegation of August 1944 Stalin had said that Communism fitted the Germans "like a saddle [fits] a cow". T.V. Volokitina *et al.*, *Sovetskii Faktor v Vostochnoi Evrope. 1944-1953*, Vol. 1, *1944-1948*, Moscow, ROSSPEN, 1999, p. 87. That did not prevent him, when the time was ripe, from pushing East Germany down the Communist road.

48 *Sochineniia*, Vol. 2, p. 304.

49 *Sochninenia*, Vol. 5, pp. 34, 46.

50 Djilas, *Conversations*, p. 114. In his discussion with the SED leadership on 4 July 1952 Stalin confirmed his faith in German capabilities: "It would be wrong to think that the Hungarians are more capable than Germans". See "Conversation between Joseph V. Stalin and SED Leadership", 4 July 1952. Library of Congress, Dmitrii Volkogonov Collection [Cold War International History Project: http://wilsoncenter.org/index.cfm?topic_id=1409&fuseaction=topics.home].

51 B.H. Bayerlein (ed.), *Georgii Dimitroff. Tagebücher 1933-1943*, Berlin, Aufbau, 2000, p. 641.

52 See C. Kennedy-Pipe, *Stalin's Cold War. Soviet Strategies in Europe, 1943 to 1956*, Manchester, Manchester UP, 1995, p. 42.

53 Mar'ina, "Peregovory", p. 15; *idem*, *Voprosy istorii* 1, 2001, p. 16.

bered being at a congress in Germany when a group of German Social Democrats had arrived late because they had felt obliged to wait to have their tickets confirmed. There would never be a German revolution "because you would have to step on the lawns". The Germans were "a queer people, like sheep. I remember from childhood: wherever the ram went, all the rest followed".[54] During the Yalta conference of February 1945, Stalin reminded Churchill of the "extraordinary discipline in the Germany of the *Kaiser*". He repeated the same story which he had told Djilas about the Germans "obediently" waiting on the platform until their tickets were checked.[55] The Germans were unique in being heroes and slaves at one and the same time.[56]

What Inspired Stalin's Views of the European Nations?

Summarising, Stalin's Europe was divided into an Eastern and Western half. The dictator measured up the two halves along two axes, civilisation and strength of character, to be more or less negatively correlated. The Eastern nations, with possible exceptions like the Latvians and Romanians, were admittedly relatively uncivilised but were endowed with impressive martial qualities. Despite their character flaws, the developed Germans and relatively developed Poles formed part of the heroic East. The Western nations (including the Americans and the Jews, who were in terms of character part of the Western world) were considered the more developed, but with their commercial orientation they suffered from a deplorable lack of iron in their character. This routine stereotyping of nations justifies the conclusion that Stalin was emotionally stamped with a powerful cultural-psychological essentialism.

It has become an object of debate in recent years whether Stalin attributed the characters of nations to their racial make-up. Francine Hirsch argues convincingly that the Stalinist scientists and policy makers did not consider biological race a determinant of psychological and cultural traits, and considered nations and nationalities altogether differ-

---

54 Djilas, *Conversations*, p. 79.

55 Nevezhin, *Zastol'nye Rechi*, p. 438. Stalin made the following note in an undated copy of Karl Kautsky's 1919 *Terrorism and Communism*. Kautsky quoted Marx to the effect that the German workers tended to follow "a saviour (like Lassalle)". Stalin commented in the margin: "the old sin". RGASPI, f. 558, op. 3, d. 91, p. 156.

56 In August-September 1911 Stalin read P. Kogan's *Ocherki po istorii zapadnoevropeiskikh literatur*, Vol. 1 (on this date see comments by his acquaintance P.B. Onufrieva, RGASPI, f. 558, op. 4, d. 647). On p. 346 the author notes that the French national spirit was classicist, logical and schematic, whereas Romanticism harmonised with the "dreamy" German character. Stalin commented "nonsense!" [*gluposti*] in the margin (f. 558, op. 1, d. 32).

ent entities from races.[57] Eric Weitz argues that, nevertheless, national characters under Stalin were not deduced from the circumstances in which particular nations found themselves, but were considered fixed givens. The nations were in other words themselves "racialised". In Weitz's definition, racism is not necessarily about skin colour. "Race is present when a defined population group is seen to have particular characteristics that are indelible, immutable, and transgenerational".[58] Weitz refers to Etienne Balibar's concept of "'culturalist' racism", and to the latter's argument that "biological or genetical naturalism is not the only means of naturalising human behaviour [...] *culture can also function like a nature*". Culture no less than race may lock individuals immutably into a genealogy.[59] The problem with this definition is that, without the racist hypothesis of heredity, it is hard to come up with another plausible mechanism by which cultures could ever become immutable and all-encompassing. It is, in other words, hard to avoid the conclusion that in fact nurture *cannot* function like nature.[60]

Stalin, for one, did acknowledge that national characters were not fixed once and for all, but were developing, and needed to be explained. To be fair, the casual comments he made in the course of the years were not unambiguous. They betray a certain fuzziness of thinking and little interest in consistency. Sometimes it seems as if national character is indeed a given rather than an *explanandum*. It is for example unclear whether in Stalin's perception the Jews had a mercantile character because they were employed in that sphere, or *vice versa*. The Germans were still the same foolhardy Teutons they had been two thousand years ago. "You can't change the Germans..."

---

57 F. Hirsch, "Race without the Practice of Racial Politics", *Slavic Review*, 61/1 (2002). See also A. Weiner, "Nothing but Certainty", *Slavic Review*, 61/1 (2002); and Francine Hirsch, *Empire of Nations. Ethnographic Knowledge and the Making of the Soviet Union*, Ithaca, Cornell UP, 2005, part 3.

58 E.D. Weitz, "Racial Politics without the Concept of Race: Reevaluating Soviet Ethnic and National Purges", *Slavic Review*, 61/1, 2002, pp. 3, 7-8.

59 E. Balibar, "Y a-t-il un 'néo-racisme'?", in E. Balibar & I. Wallerstein, *Race, nation, classe. Les identités ambiguës*, Paris, Éditions La Découverte, 1988, pp. 34, 37.

60 Weitz furthermore refers to George Fredrickson, who argues that racism can be said to exist in all cases when ethnic groups assert their superiority *vis-à-vis* other groups who are thought "because of defective ancestry, to possess a set of socially relevant characteristics that disqualify them": G.M. Fredrickson, *The Comparative Imagination. On the History of Racism, Nationalism, and Social Movements*, Berkeley, University of California Press, 1997, p. 85. Because ancestry remains central to it, in Frederickson's model the link with biological heredity is in fact not completely severed. Race is only expanded to embrace any ethnos with supposedly hereditary psycho-cultural traits.

Again, Stalin did on occasion indeed suggest that race underlay the stability of popular character. In 1944 he told a Polish delegation that the Slavic peoples were close to each other in "*kin, blood,* language, character, deep humanity and an understanding of the idea of progress", but he did not elaborate on this.[61] When in November 1947 the French Communist leader Maurice Thorez told him that the Belgian Communists were stronger in the French speaking regions than in the Flemish parts of the country, Stalin casually remarked "that in the racial respect the Flemish are closer to the Germans". What he intended to suggest with his remark, if anything, remained unclear.[62]

There is no doubt that Stalin believed in the essentialism of national character. In his 1913 fundamental text about the nation, he defined it as "neither racial nor tribal", but as an essentially modern formation. The nation was a "*historically formed, stable community of people*" which arose on the basis of a "*community of language, territory, economic life and psychological make-up, manifesting itself in a community of culture*". So, without such a thing as national character, a unique psychology of their own, a community could not even be considered a nation to begin with. What is more, national characters were shared by all members of the nation; they formed a "mentality [*dykhovnyi oblik*] of the people who come together in a nation". This "stable" mentality was furthermore transmitted over time, as a "psychological make-up [*psikhicheskii sklad*] that was formed among them from generation to generation as a result of identical conditions of existence".[63]

Nonetheless, for all its comprehensiveness and stability, national character was not racially determined, but "a reflection of the conditions of life" of a nation. Stalin described it as a mere "lump of impressions received from the surrounding environment". Consequently, national character did in his view "not constitute something given once and for all, but changes together with the conditions of life". The main example he discussed was that of the Jews. Their diasporic condition and their living a common economic life with the nations among whom they settled would unavoidably put a mark on their national character, which could not be preserved unchanged.[64] Two decades later Stalin explained how habits were formed, taking a primitive position in the nurture/nature debate, on the nurture side. In May 1933 he enlightened a foreign visitor as follows:

---

61 RGASPI, f. 558, op. 4, d. 612, ll. 4-6 (my italics).

62 "Zapis' besedy tov I.V. Stalina s sekretarem TsK Frantsuzskoi kompartii Torezom", *Istoricheskii arkhiv*, 1, 1996, p. 11.

63 *Sochineniia*, Vol. 2, pp. 293, 296, 301.

64 *Sochineniia*, Vol. 2, pp. 296, 299-300.

> I believe you shouldn't put the question as if the workers of some nation would be incapable of mastering new technology. You can look at the matter from a racist point of view, like in America for instance, where the negroes are considered the "lowest people" – but they master technology no worse than whites. [This] is no biological question, no question of heredity, but a question of time: today they have not mastered [technology], tomorrow they will learn it and master it.[65]

Stalin often repeated that Marxism explains human phenomena from the point of view of sociology, not biology.[66]

After the Great Patriotic War, the Soviet dictator began to emphasise the uniqueness of national characters more than ever. In October 1945 he told members of the Finland-USSR society that "each people, small as well as large, has its worth and makes its contribution to the historical development of humanity".[67] In the course of the years this was turned into a solemn formula, repeated ritually to visiting foreign delegations. Each nation had "certain particularities characteristic only to it", which represented the "contribution which each nation makes to the common treasure-house of world culture".[68] At the same time, though, in the dictator's last years the environmental basis of national characters was even more strongly emphasised than before. By putting his weight behind Trophy Lysenko's neo-Lamarckian biology, he supported a theory that underscored the malleability of human heredity. The idea of national communities with fixed psychological characteristics was decried as racist.[69]

---

65 *Sochineniia*, Vol. 13, pp. 264-265.

66 For example, remarks at a meeting of the Military Council of the Commissariat of Defence in June 1937: I.G. Murin, "'Nevol'niki v rukakh germanskogo Reikhsvera': Rech' I.V. Stalina v Narkomate oborony", *Istochnik*, 3, 1994, p. 74. In November 1944 Stalin noted that "Soviet patriotism" did not have "racial or nationalist prejudices for its basis". *Sochineniia*, Vol. 2 [XV], p. 161. See also his remarks to Romanian party leader Gheorghe Gheorghiu-Dej in 1947 (Volokitina *et al.*, *Vostochnaia Evropa*, pp. 565, 582. For other occasions when Stalin expressed himself against racism: Nevezhin, *Zastol'nye Rechi*, pp. 258, 332.

67 RGASPI, f. 558, op. 1, d. 5379, l.1.

68 See for two occasions in 1948: Nevezhin, *Zastol'nye Rechi*, pp. 503, 508. Perhaps the first systematic discussion of Stalin's views of nations as the individualities of humanity is to be found in Mehnert, *Stalin versus Marx*, p. 29ff.

69 Nikolai Krementsov (*Stalinist Science*, Princeton, Princeton UP, 1997, pp. 204-207, 260-275) discusses Soviet attempts to reformulate pedagogy and human physiology in Lysenkoist terms. For Lysenkoism and racism see also Y. Slezkine, *Arctic Mirrors. Russia and the Small Peoples of the North*, Ithaca, Cornell UP, 1994, p. 314. Robert Tucker, *The Soviet Political Mind. Stalinism and Post-Stalin Change,* revised ed., London, George Allen & Unwin, 1972, pp. 143, 151ff., discusses Stalinist "transformism" and the uses of Pavlovian physiology to underscore the idea that humans were infinitely malleable through the "conditioned reflex". See also I. Halfin,

If the differences between the West and the East European nations were not of a racialist nature, the question then arises of what the particular "conditions of life" were that gave rise to these differences. To my knowledge, Stalin never systematically discussed this, but the drift of his thinking is clear enough. As he saw it, the capitalist system inevitably produced a commercialist mentality that prioritised financial motives and thereby undermined the popular character. The mechanism operated on the level of individuals, too. Solomon Volkov quotes a typical anecdote of the leader, who at the 1933 All-Union Music Performance Competitions handed over a large cash prize to an eleven-year-old boy, and commented jokingly: "Well, Busya, now you're a capitalist and probably will become so conceited that you won't want to invite me to visit you".[70] In a November 1947 conversation with French Communist leader Thorez he said, "when a person gets dollars his head will get empty. A full pocket makes for an empty head".[71] The delayed socio-economic development of Eastern Europe happily saved it from the psychological degradation which the advances of the commercialist civilisation brought on the West.

To this one must add Stalin's personal insights into human psychology as another powerful element. When he noted that British and French resolve had been weakened due to the overconfidence generated by their victory in the Great War, he was reiterating a general truth that guided him throughout his life: success tends to make people rest on their laurels. The matter points to a deep paradox in Stalin's thinking. As a historical optimist, he felt that all nations had a holy obligation to develop powerful modern states. But success tended to make people overconfident. To be sure, Stalin did acknowledge that decadence was not development's inevitable product. Under Communist "conditions of life", demoralisation was avoidable. Nevertheless, the developed, "satiated" nations of the West would have been much more at risk of developing a flabby mentality and a faltering of the will than the East Europeans, who could not afford complacency.

But it would, finally, be mistaken to assume that Stalin's interest in the matter of the character of European nations was primarily of an intellectual nature, or that he was only driven by ingrained prejudices – Marxist, racist or other. His remarks were also inspired by strategic considerations and by his understanding of politics as mobilisation. The

Terror in my Soul. Communist Autobiographies on Trial, Cambridge Mass., Harvard UP, 2003, p. 238ff.

70 S. Volkov, *Shostakovich and Stalin. The Extraordinary Relationship between the Great Composer and the Brutal Dictator*, London, Little, Brown, 2004, p. 131.

71 "Zapis' besedy tov I.V. Stalina s sekretarem", 1996, p. 17.

circumstances in which he made his remarks more often than not suggested a political agenda. His good words about the Yugoslavs, Poles, Finns and Albanians were directed mostly to politicians and delegations from these countries. It was his way of providing his allies with confidence in their own strength. By describing the Soviet people and their allies as heroes, he hoped to instil a fighting spirit into them. His disdainful remarks about Western decadence were made mainly in discussions with representatives of the Soviet public, and with allied politicians from Eastern Europe and People's China. Describing the capitalist nations as flabby merchants served the dual purpose of providing his own people with confidence, and of rooting out the decadent mentality that would hamper their fighting abilities.

Stalin's view of Europe may perhaps be most fruitfully analysed as a screen on which he projected his heroic ideals. His remarks about the European nations not only reflected what he thought these nations were, but also what he wanted them to become. As leader of the USSR and then of the socialist camp, he hoped to see the West morally weakened and therefore easy to defeat. Conversely, by convincing the Soviet people and their allies that they were of heroic stock, he attempted to make them adopt the heroic mentality he considered an essential condition for attaining victory.

Heroic Modernity

This returns us to the question of the similarities and differences between Stalinist and Conservative-Revolutionary heroism. In contrast to the Stalinists, the Conservative Revolutionaries never became more than an ideological movement. Though much of their programme and mentality was adopted by the Nazis, the Conservative Revolutionaries themselves never got the chance to lead a state and show what they were worth. Their ideas always preserved a desperate, outrageous quality that was lacking in the Stalinist case.

The Conservative Revolutionary ethos was fundamentally anti-economic. Society should be liberated from the dominance of economic concerns forced upon it by liberals and Marxists. War, an honourable undertaking and a school of character-building and sacrifice, should be the main pursuit of the new society. War and battle were, moreover, no instruments of a greater goal but goals in themselves, the occupation of the hero, who was out to confirm his own virile vitality and for whom victory and defeat were of secondary importance. As Romantics, the Conservative Revolutionaries had initially been disposed against modern technology. They tended to see the machine as an evil force sapping the independent will of the hero and condemning humanity to mediocrity and indolence. However, important ideologues such as Sombart and Spengler and most emphatically Jünger reached the conclusion that the irrational and destructive principles inherent in modern technology were precisely what the hero needed to enhance his powers. Without it, Germany would not stand a chance in the coming war of revenge they eagerly awaited.[72]

With their anti-egalitarian ethos and authoritarianism, the Conservative Revolutionaries drew their inspiration from archaic principles that had been overturned by the French Revolution. In part, the movement was therefore a fundamentally anti-modern one. But most specialists agree that it was, in other ways, modern in inspiration.[73] Jeffrey Herf emphasises the anti-rationalist and anti-Enlightenment goals of the movement, but also that by their "irrationalist embrace of technology", rearmament and industrial rationalisation, the "reactionary modernists" partook importantly in modernity.[74] War as an activity for its own sake, serving the gratification of the hero even without the necessary prospect of victory, has also been analysed as expressing a typically modern mentality. Rolf Sieferle calls the Conservative Revolution a Romantic "lifestyle-revolt", aiming fundamentally for self-realisation.[75]

In sharp contrast to the German revolutionaries, the Stalinist heroic ideal was economically oriented. As we saw, Stalin highly appreciated the martial spirit and combative qualities, but labour heroism was no less important. He did not reject commercialism as part of a decadent materialist orientation, but as a decadent deformation of the essentially healthy pursuit of material prosperity. As long as labour discipline was not weakened by comfort-seeking it was perfectly legitimate for the hero to seek a well-off existence. Constructing an economically developed state was one of Stalin's main purposes. Jeffrey Herf rightly notes that

---

72 For the primacy of war over economics, see Herf, *Reactionary Modernism*, pp. 57-58, 103, 147. For the idea of war and battle as goals in themselves, and its connection with Nietzsche's understanding of history as cyclical and eternally recurring: Mohler, *Die Konservative Revolution*, pp. 106ff, 157-161; Herf, *Reactionary Modernism*, pp. 21, 27; Sieferle, *Die Konservative Revolution*, pp. 36-39, 137, 142; Woods, *The Conservative Revolution*, pp. 29-49. On the Conservative Revolution and technology, see Breuer, *Anatomie*, pp. 70-78; Herf, *Reactionary Modernism*; and Sieferle, *Die Konservative Revolution.*

73 See Herf, *Reactionary Modernism*; Breuer, *Anatomie*; Sieferle, *Die Konservative Revolution*. Mohler (*Die Konservative Revolution*, pp. 18-20, 146-151) remarked that the Conservative Revolutionaries objected to being identified with "reactionaries" who wanted to restore the *ancien régime*.

74 Herf, *Reactionary Modernism*, pp. ix, 1-3.

75 Sieferle, *Die Konservative Revolution*, p. 38. See also Herf, *Reactionary Modernism*, pp. 15, 27, on the Romantic "celebration of the self" as a typically modern phenomenon.

"Jünger's cult of labor, then, is not parallel to the productivism of the Soviet Union. Rather it stems from the affinities he saw between peacetime labor and military obedience".[76] Herf adds that the Stalinists were "highly enthusiastic about science and technology without the residue of philosophical irrationalism" that burdened the revolutionaries of the Right.[77]

The Stalinist conception of heroic warfare also differed sharply from the Conservative Revolutionary model. Following Nietzsche's historical pessimism, the revolutionaries of the Right were deprived of the certainty of victory and tended to see battle as a gruesome game.[78] But the Marxist-Leninist historical laws guaranteed the victory of the Communist cause. The Soviet warrior-hero should spend his energies efficiently, bowing to the demands of the rational rules of engagement, in order to accelerate victory. Fights should only be engaged in if the chances of success were good. When the pilot Valerii Chkalov pledged in March 1938 that he would die for Stalin, the leader reacted sarcastically, "dying is something every fool is capable of". Instead of fulfilling the "empty task" of dying one should destroy the enemy.[79] Typically, in February 1948 Stalin told Yugoslav and Bulgarian party leaders, who hoped to continue support for the hopeless Greek guerrilla struggle:

> You are under the impression of a "moral obligation". If you cannot lift the weight which you have hoisted upon yourself, you must admit it. [...] We do not have such categorical imperatives. The entire question rests in the balance of forces. We go into battle not when the enemy wants us to, but when it's in our interests.[80]

We might call the typical Stalinist hero a "Scythian engineer". His ethos was more productivist and rationalist than that of his right-wing counterpart. In the case of the heroes of the Right, modern technology and the economic machinery of the state were the instruments the hero hoped to recruit to enable him to live out his vitalistic fantasies. In the Stalinist case, the order was reversed; the Romantic heroism Stalin hoped to impose on the Soviet people was the means to an end. Ferocity and ruthlessness were mobilised to force economic and military breakthroughs and to provide the Communist state with victory. The instrumental role of heroic emotion is reflected in Stalin's own words, when

76 Herf, *Reactionary Modernism*, p. 103.

77 *Ibid.*, p. 223.

78 Mohler, *Die Konservative Revolution*, p. 160, calls Nietzsche's *amor fati* the "essence of the Conservative Revolutionary attitude". See also Sieferle, *Die Konservative Revolution*, p. 142.

79 Nevezhin, *Zastol'nye Rechi*, pp. 192-193.

80 Gibianskii, "The Soviet Bloc", pp. 130, 133.

he remarked in January 1929, for example, that "every period in a national development knows its pathos. In Russia we now have a pathos of construction".[81] David Priestland quotes him to the effect that the "idealisation" of man and of the new social system could be used to create "the sort of romanticism that would move us forward".[82]

In these different perspectives we perhaps find a reflection of the historical circumstances in which the two hero cults arose. In the German case, there was no economic backwardness to overcome but a preoccupation with war. The lost war was the primary national trauma. And the fact that it would be very difficult to wage another, victorious war against the same, powerful enemy states provided the dreams of these nationalists with their desperate and irresponsible edge. In the Soviet case, the war trauma was considerably less sharp. The Bolsheviks had not identified with the Tsarist government; its unsteady performance in the Great War did not weigh on their conscience. Their war had been the Civil War, which they had won. Their first preoccupation was the essentially economic feat to be achieved of "constructing Socialism", in order to bring Soviet Russia up to the level of the advanced nations.

Yet the fact that Conservative Revolutionaries and Stalinists shared cults of the heroic mentality and an abhorrence of commercialism and comfort-seeking remains striking. It seems that the defeat which the Western powers brought upon Germany in the First World War and Russian-Soviet economic backwardness produced similar reactions. In both cases the goal was formulated to get even with the West – by preparing for another, this time hopefully victorious war; or by constructing an economically and militarily powerful modern state. In both cases it was felt that these modernising goals could not be achieved without extraordinary mobilisational means. The heroic mentality, which would prevent the nation from wasting its energies in empty comfort-seeking, was assumed to be the crucial condition to advance from the disadvantaged position one felt oneself to be in. Seen in this light, both movements expressed a spirit of what we might call "heroic modernity".

---

[81] *Sochineniia*, Vol. 13, p. 149. "Pathos" was for Stalin also an important criterion to judge the value of nations. In a speech on the *Kratkii kurs* in September 1938 he noted that German history was "very poor". The Germans had some "moments of pathos" but it didn't compare to Russian history. "I.V. Stalin v rabote nad 'Kratkim kursom istorii VKP(b)'. Okonchanie", *Voprosy istorii*, 4, 2003, p. 5.

[82] D. Priestland, "Stalin as Bolshevik Romantic: Ideology and Mobilisation, 1917-1919", in S. Davies & J. Harris (eds.), *Stalin. A New History*, Cambridge, Cambridge UP, 2005, p. 194.

Conclusion

To conclude, dictator Joseph Stalin saw it as his life's purpose to turn Soviet Russia into an economically and militarily powerful, modern state. From his point of view, Russians and other East European nations were only half-civilised, making up a midway station between Asia and Europe proper. At the same time, Stalin did, like the German Conservative Revolutionaries, divide up the European nations into two fundamentally different cultural-psychological communities of heroic and commercialist frames of mind. The less developed East took the lead over the West in terms of character and stamina. What Germany was for the Conservative Revolutionaries, Soviet Russia was for Stalin: the beacon of heroism.

In taking this view, Stalin was not inspired by straightforward racism. Though he was driven by a profound and primitive national essentialism, he did, somewhat paradoxically, acknowledge that national character remained rooted in socio-economic and other circumstances. And it was precisely the lack of healthy capitalist development, and of development in general, that had preserved the heroic spirit in the eastern parts of the European continent. But Stalin's ascription of courage and cowardice to respectively Eastern and Western Europe should above all be interpreted in political terms. It served the dictator's mobilising purpose of equipping the Soviet people and their allies with the ferocious spirit he believed to be essential for the military and economic competition with the imperialist West and for the "construction of Socialism".

Finally, I would interpret the remarkable similarities between Stalinist and Conservative Revolutionary heroism not primarily in terms of their ideological-intellectual backgrounds. Common roots in early twentieth-century Romantic anti-capitalism form an unmistakable factor.[83] But the main common denominator was the common political intent to educate nations in a heroic spirit of sacrifice and of the discard-

---

83 Bernice Rosenthal (who edited *Nietzsche in Russia*, Princeton, Princeton UP, 1986; and *New Myth, New World. From Nietzsche to Stalinism*, University Park Pa., Pennsylvania State UP, 2002) and Katerina Clark (*Petersburg. Crucible of Cultural Revolution*, Cambridge Mass., Harvard UP, 1995) argue that Stalinist heroism originated as part of the same Europe-wide *fin de siècle* phenomenon that gave rise to the Conservative Revolution. "Romantic anti-capitalism" and Nietzschean influences are discussed. See also G.L. Kline, *Religious and Anti-religious Thought in Russia. The Weil Lectures*, Chicago, University of Chicago Press, 1968, pp. 103-119. Hans Günther, *Der sozialistische Übermensch. M. Gor'kij und der sowjetische Heldenmythos*, Stuttgart, Metzler, 1993, observes a continuity running from Maksim Gor'kii's early works, with their contrasting of the heroic to the petit bourgeois, slavish and decadent culture, to the later Stalinist hero cult.

ing of all frivolity. Only thus, Conservative Revolutionaries and Stalinists were at one in believing, would Germans or Russians stand a chance of winning the military and economic competition with the predominant capitalist West.

CHAPTER 3

The Nazi New Order and Europe

David BARNOUW

The triangular castle of Wewelsburg is located in Paderborn in North Rhine-Westphalia, where it has stood since the beginning of the seventeenth century. In 1934 *Reichsführer* Heinrich Himmler (1900-1945) leased the castle for 100 years with the intention of recasting it into a training school for high-ranking SS officers. Beneath the great tower, the *Nordturm*, a marble mosaic was installed in the shape of a dark sundial. This space was to be used by a maximum of thirteen men: twelve SS officers and Heinrich Himmler himself, the Nazis' twentieth-century rendition of King Arthur's Knights of the Round Table. Below, a vault was sunk halfway into the ground, decorated with swastikas in the ceiling. The *Nordturm* would serve as the *Mittelpunkt der Welt*, the centre of the world. The renovation, which took place between 1939 and 1943, was for the most part accomplished with forced labour: political prisoners from concentration camps and POWs. This castle symbolises, as it were, the long-term thinking of the Nazi elite – in this case of the SS – who aspired to play a major role in Europe's New Order, or *Neuordnung*. The *Mittelpunkt der Welt* is where *Ostforschung* and *Westforschung* research converged, buttressed by geopolitics. The sundial and the cult of the Round Table represent the – frequently overestimated – occult side of SS ideology, the use of forced labour, and the contempt for non-Teutons. At the end of the war there was an attempt to blow up the castle to prevent it from falling into Allied hands – a fitting coda to the ephemerality of the Thousand-Year *Reich*. The Wewelsburg has since been renovated and houses both a youth hostel and an exhibition about the "evil" SS.

It goes without saying that the most important decisions in Nazi Germany were not made in Paderborn, but in Berlin or at the various military headquarters where the Führer happened to be during the war. Nonetheless, the bet was on a Thousand-Year *Reich*, a *Reich* in which the *Ordenburg Wewelsburg* was certain to play a role. The anticipated New Order of Europe, to be achieved with or without force, by means of diplomacy or with soldiers, was backed by numerous scholars.

Under the terms of the Treaty of Versailles that came upon the heels of her defeat in World War I, Germany lost more than 10 per cent of her territories and was saddled with a huge reparations bill. Most Germans agreed that the country should try to recover some of the lost territories. The Nazis, however, sought a radical *Neuordnung* of Europe, and to avoid another humiliation like Versailles by making Germany as strong and as powerful as possible. Hundreds of scholars were mobilised to establish a scientific foundation for this craving for land – *Lebensraum* – placing their academic training in the service of an expansionist and (for the most part) racist policy. Geopolitics, *Ostforschung* and *Westforschung* paved the way for the conquest of Europe: to give Germany and the Germanic peoples their historic due as masters over inferior peoples. This chapter surveys the different scholarly currents involved in shaping the new Europe, as perceived by the Nazis.

By way of preface, one might mention the discussion concerning the "Nordic race", which was by no means confined to German or Nazi scholars. Since the end of the nineteenth century and particularly since the beginning of the twentieth, there had been attempts to move towards the perfection of humankind by means of the science of ethnology, or more especially, eugenics. Although it might seem to be a discipline ideally suited to the Nazis, it was of course taken perfectly seriously and indeed embraced by large parts of the academic community on both sides of the Atlantic.[1] Eugenics was concerned with the directions in which the human race was moving, and how that process might be most valuably assisted by science.

At the same time there was the question of the *origins* of the human race, and especially those of the most advanced part of it: where did the "Nordic race" originate, and who were the forefathers of the Aryans? In these matters the study of language took a central position in anthropological research, with linguistic comparison playing a principal role. Research apparently revealed that the Indo-Iranians had spoken Indo-European languages. In Old Persian the word *Ariya* means "noble", thought to have derived from *Ãrya* in Sanskrit, which was spoken on the Indian subcontinent between about 1200 and 500 BC. The step from language to race was of course taken far too readily, with the result that the Nordic Aryans were held to be descended from Asian Sanskrit speakers. The term "noble" was appropriately welcome: the Nordic Aryans belonged to a "higher" or better "race". "Through the lectures

1 For a survey of the literature, see D. van Galen Last, "Rassenhygiëne: de dodelijke omhelzing van wetenschap en nationaal-socialisme. Literatuuroverzicht", in *Volkseigen, ras, cultuur en wetenschap in Nederland 1900-1950; Elfde Jaarboek Oorlogsdocumentatie*, Zutphen, Walburg, 2000, pp. 213-247.

and books of the great philologist Max Müller (1823-1900) in Oxford and Ernest Renan (1823-1892) in France, the Aryan myth had become established dogma throughout European learned society by about 1860".[2] Right up to the time that Hitler came to power, it was a theory widely accepted by many scientists, whether German, French or Anglo-Saxon. In countries where white superiority had long been promulgated, such as the United States, the Aryan theories provided a degree of scientific support for such doctrine, employed not only against blacks, but against Italians and Irish too. With the fall of the Third *Reich*, all public political deployment of eugenics ceased, and to think in terms of "higher" and "lower" races became taboo.

Geopolitics, *Ostforschung* and *Westforschung* have a common objective: to establish the foundations for the German drive to the east and the west on a scientific, often quasi-scientific, basis. This discipline, dubbed *Ostforschung*, frequently was steeped in racism. Blood and soil, archaeology, pre-history, medieval history, ethnology, genealogy, religion, and above all racial science (whatever that might be): all of it was harnassed to its pursuit.

Many of the scholars who laid the theoretical foundations for racial imperialism and mass murder before the outbreak of war unstintingly served the Nazi regime after it came to power in 1933, and sometimes saw their ideas come to fruition during the war itself. After the war they returned to the safety of their universities, denied that their scholarship had anything to do with mass murder, and resumed their work. Much of the research they had conducted in occupied territories turned up in scholarly forums after the war, with no questions asked.

"From Genealogy to Genocide and back to Genealogy"[3] was how Eric Schmaltz and Samuel Sinner summed up the careers of several of these scholars. Here the culprits were two ethnic Germans, originally from the Ukraine, who rose to prominence after the war not only in German-Ukrainian organisations in Germany such as the *Landmansschaft*, but also in the USA, where they were associated with the *American Historical Society of Germans from Russia* and *Germans from Russia Heritage Society*, which saw fit to honour these two ex-Nazis as diligent genealogists.

To talk about *Ost-* and *Westforschung* in Germany is to talk about scholars who claimed to have pursued pure and value-free scholarship. True, their work had been abused by the Nazis, but that was all. Having

2 N. Goodrick-Clarke, *Hitler's Priestess: Savitri Devi, the Hindu-Aryan Myth and Neo-Nazism*, New York, New York UP, 1998, p. 33.

3 In I. Haar & M. Falbusch (eds.), *German Scholars and Ethnic Cleansing, 1920-1945*, New York, 2005, pp. 51-85.

been briefly quarantined, they returned to their institutes, kept quiet about the war, like most of the others who returned after emigration. This, of course, was not unique to historians.

Marburg has been the centre of *Ostforschung* since 1950. That was where the *Johann Gottfried Her-Forschungsrat* was founded, to be followed, two years later, by the *Zeitschrift für Ostforschung*. Precisely because of the question of the *Vertriebenen* from former German territories, there could be a continuity in *Ostforschung*. Humboldt Universität in Berlin (DDR) had an "Abteilung für Geschichte der imperialistischen Ostforschung", and published such titles as *Informationen über die Geschichte der imperialistischen Ostforschung*. The Cold War made sure that any GDR criticism of *Ostforschung* could be shunted aside. Only in 1994 did the Federal Republic stop using the concept of *Ostforschung*: *Ostmitteleuropaforschung* seemed a more neutral designation.

The *Deutscher Historikertag* in Frankfurt in 1998 marked a belated breakthrough in the discussion about this thorny subject, raising the problem of continuity and the political dimension of historical writing. This was due, possibly, to the generation gap, as evidenced by the following locution in the introduction to a collection based on the *Historikertag*: "Guilty parents, easygoing sons, and stern grandsons".[4] It seemed as if it were only in Germany were historians prepared to come to terms with the *Ost-* and *Westforschung* of their ancestors. Fortunately, this has changed. In the autumn of 2003 a symposium took place in the Netherlands on the topic "De Westforschung en Nederland", and the contributions were incorporated in a thematic edition of the *Tijdschrift voor Geschiedenis*.[5] In the discussion, the question of whether the *Westforschung* was as bad and, in a certain sense, as criminal, as the *Ostforschung* hovered in the background in the Netherlands as well. Yet a full-blown discussion was still missing. The Dutch historian Chris Lorenz attributed this absence to two factors: first, the Dutch still viewed themselves solely as victims of the Second World War, and second, the Dutch *poldermodel* – the predilection for consensus decision-making – tends to discourage intensive debate.[6]

A few years later, the *Westforschung* debate re-emerged, this time in a report about the "war past" of ethnologist P.J. Meertens (1899-1985), the father of folklore research in the Netherlands. The Meertens Instituut voor Dialectologie, Volkskunde en Naamkunde (it researches and

4 W. Schulze & O.G. Oexle (eds.), *Deutsche Historiker im Nationalsozialismus*, Frankfurt am Main, 1999, p. 27.

5 *Tijdschrift voor Geschiedenis*, 118/2, 2005.

6 Chris Lorenz, "De Westforschung – een geval apart?", *Tijdschrift voor Geschiedenis*, 118/2, 2005, pp. 252-271.

documents the diversity of language and culture in the Netherlands) was named after him. An objection was raised to the institute's name, considering it tarnished. A Committee of Three investigated the matter and concluded that Meertens was "an opportunistic advocate of Folklore," but not "a 'collaborator' in the usual sense of the word".[7] Since the Second World War "the usual sense of the word" has been liberally applied, but that is beside the point. In the section "Volkskunde en Westforschung," the term *Westforschung* was rather depoliticised. The ruling on folklore went: "Folklore could not and should not be dismissed out of hand as Nazi scholarship".[8] The institute was allowed to continue to bear its name and the discussion quickly petered out.

But not in Germany. In his article, "Hans Rothfels: Issues and Paradoxes of an International Debate",[9] the American historian John L. Harvey convincingly demonstrated that the debate has not come to an end and that interpretation and re-interpretation are ongoing. Nor is the debate confined solely to the Nazi period. Both Weimar and the Federal Republic have become part of it as well. Hans Rothfels (1891-1976) is a case in point. Appointed Professor of Early Modern History in Königsberg (now Kaliningrad) in 1926, Rothfels was as enamoured with the *Drang nach Osten* as he was unenamoured with the Poles and other East Europeans. But it is not easy to pin him down. Was he conservative, right-radical, or an outright Nazi? And what was his influence on his students who actually marched eastward with Hitler's troops? Paradoxically, he was unable to retain his professorship under the Nazi regime: his Jewish background forced him to emigrate to the USA. However, he returned to Germany in the 1950s and taught in Tübingen. He would remain a conservative go-between for the US and Germany. But how far did his pre-war *Volksgeschichte* correspond to his post-war *Zeitgeschichte*? "It points to a difficulty with situating Rothfels within clear paradigms, either as *Vernunftrepublikaner* or fascist", sighed Harvey at the end of his article.

Before delving into geopolitics, *Westforschung* and *Ostforschung*, we should take note of the following three points. First, quite apart from ideology, nearly all Germans agreed that Germany had been treated scandalously at Versailles, and many, if not an overwhelming majority, regarded a return to the pre-1914 borders as an absolute necessity. Second, the plans often seemed to be far-reaching indeed: one German *Reich* from the Atlantic to the Urals. It would be helpful to keep in

7 *Bevindingen over P. J. Meertens op grond van literatuur en geraadpleegde bronnen* (Rapport van de Commissie van Drie), Amsterdam, 2006.

8 *Bevindingen*, p. 24.

9 *Sozial Geschichte*, 22/1, 2007, pp. 7-39.

mind, however, that we are dealing with a Thousand-Year *Reich*, and that not all plans had to be fully realised in a single manoeuvre. This also applied, for example, to the megalomaniac projects of Hitler's favourite architect, Albert Speer. Third, irrespective of what was being imagined and depicted, and irrespective of what the Nazi party, the generals, Foreign Affairs, or Himmler's *Ostministerium* might say, in the final analysis it was Hitler himself who made the important decisions. And interestingly, while on one level certain principles were considered sacrosanct – among them the centrality of race and the belief that Bolshevism must be destroyed – on another Hitler could be extraordinarily flexible and opportunistic, the best examples being the Molotov-Ribbentrop pact and the numerous treaties he had concluded before that.

At the end of 1940 Nazi Germany appeared to be at the height of its achievements: the Saar-region was German again, as was the Rhineland; Austria was "*Heim ins Reich*", the Sudetenland was part of Germany, Bohemia and Moravia had become German protectorates, Memelland was "wieder Deutsch", and the western part of Poland was occupied, as were Denmark, Norway, the Netherlands, Luxembourg and a large part of France. It came very close to living up to the first part of the German National Anthem, the *Deutschlandlied*:

Von der Maas bis an die Memel
Von der Etsch[10] bis an der Belt
Deutschland, Deutschland über alles
Über alles in der welt.

And all this had been achieved with a minimum of war casualties. It seemed that with the Soviet Union still neutral and Great Britain isolated, it was only a matter of time before Britain was defeated and Germany would be master of the European continent. But Britain was not defeated, because Hitler did not have the will to conquer or destroy the island, and he did not have the resources in the air or elsewhere to do it. His attitude towards England had always been ambivalent, with a simultaneous deep mistrust for the "Jewish democracy" and an admiration for the British Empire.[11] He wanted to use the empire model for the vast Soviet Empire, after it had been crushed by "Operation Barbarossa". Along the Ural River a "live" barrier consisting of German settlers/soldier-farmers would be required, to keep the Asians in the East. In this way a German colony would be created in Russia by means

[10] A river in the north of Italy.

[11] "Hitler's model for domination and exploitation remained the British Empire". Ian Kershaw, *Hitler 1936-1945. Nemesis*, London, 2000, p. 401.

of a feudal settlement between the eastern part of Germany and the Ural River.

The British Empire model did not fit for occupied Western Europe, so serious discussions about the future there were postponed for "after the war". But one could observe the differences in occupation policy: on the one hand Poland was brutally ruled by ruthless satraps, while at the other end of the spectrum Denmark and the Netherlands were governed with much more respect. Belgium and France had military governments, but the Netherlands was governed by civil servants. The New European Order had to be National Socialist, but how it would be implemented was a question not yet resolved. Nothing was settled, and in the meantime German Party officials, SS officers and military agencies were endlessly vying with each other to get a good share of the occupied territories.

Geopolitics

Geopolitics strives to achieve political objectives on the basis of arguments grounded in geography.[12] Geopolitics had its greatest impact in German-speaking areas, where it appealed to conservatives and Nazis in particular. Although its appeal was directly related to the German defeat of 1918, it certainly did not come out of the blue. Further, it was an elaboration of ideas that had been current for centuries, regarding the organic origins of states, and it probably was no accident that this elaboration was given its most vigorous expression in Germany. German unity was forged in the victorious war with France in 1871, the first time Germany had existed in this form, but that was not enough. Some dreamed, others wrote, of a single Greater Germany, with all German speakers under a single roof. Germany was too small for a growing population and, unlike England, France, Belgium or the Netherlands, had few colonies. She had simply arrived too late on the world stage in the nineteenth century, and had been relieved after the First World War of the surprising number she had accumulated. There was, it is true, a Ministry of Colonial Affairs, and Germany managed to acquire four African colonies, whose reputed magnificence was always kept in the public eye back home. After the Treaty of Brest-Litovsk in March 1918 toward the end of the First World War it appeared that, with the addition of the Baltic states of Estonia, Latvia and Lithuania, Russian Poland, and the Ukraine, Germany had expanded a great deal eastward. The joy, alas, was short-lived, because with the collapse of the western front nine months later, Germany surrendered to the Allies. Besides yielding

12 See G. Bakker, *Duitse geopolitiek 1919-1945: een imperialistiche ideologie*, Assen, 1967.

all these territories, Germany had to surrender yet more land, and was also deprived of her colonies.

Germany seemed have reconciled herself to being a continental rather than a colonial empire. Of course there still were German settlers in the former colonies, and there was a notable shift in how these settlers perceived themselves. Until the First World War colonial films stressed the modernisation and industrialisation introduced by the Germans. After that, the emphasis was on the *Heimat*: German language, schools, church, and superior culture in foreign climes.[13] The emphasis of colonial propaganda between the wars was on portraying Great Britain as a brutal colonial power, as opposed to peace-loving and culture-bearing Germany. That was the message, for example, of *Carl Peters, ein deutsches Schicksal (Carl Peters, A German Fate)*, a 1941 film about German Africa. The notion of "perfidious Albion" also played well in the Netherlands. This was a product of the South African Boer War in which their "kinsmen", the Boers, had suffered defeat at the hands of the British. The use of internment camps for women and children gave England a bad press. This played a significant role in Dutch public opinion during the First World War, as well as in the Second, when the Germans tried to capitalise on such sentiments.[14]

The fleet Germany constructed in violation of the provisions of the Treaty of Versailles was not designed to ply remote waters or for purposes of colonisation; rather, it was to prevent the British fleet from imposing a blockade from the sea. Colonies outside Europe played a minor role in Nazi ideology: the gaze was fixed on the East. She may have had no colonies, but Germany had no intention of reconciling herself to the Versailles *Diktat* and its imposed borders – why should these borders be forever? States, and certainly the German state, needed *Lebensraum*. Space, and the expansion of space, was a question of life and death – why allow legal hair-splitting to stand in the way? And the state with the best "racial qualities" was the most likely to succeed. In Europe that could be no other than Germany, which would not only encompass German-speaking areas, but also whatever else was needed for *Lebensraum*:

> Geography accounted for one fourth of the factors vital to the organism of state. The other three quarters were shaped by demographic, economic and

---

13 Wolfgang Führman in his lecture, "From Modern Colony to Colonial Heimat", at the conference at the University of Amsterdam, "Media and Imperialism", July 2007.

14 See I. Tames, *"Oorlog voor onze gedachten": oorlog, neutraliteit en identiteit in het Nederlandse publieke debat, 1914-1918*, Hilversum, 2006, especially Chapter 2: "De stellingen worden betrokken: polarisatie, propaganda en vredesverlangen", pp. 58-98.

military circumstances and, above all, by the qualities of the race. The whole thing could be summarised in the phrase "Blut und Boden".[15]

Through graphics and maps Germans were informed that their country's population was almost three times greater than that of south-east Europe and seventeen times that of the Soviet Union. The message could not have been clearer: there was plenty of space for land-starved Germans. There was no real reason for states like Estonia, Latvia and Lithuania to exist.

The Swede Rudolf Kjellén (1864-1922) coined the term geopolitics, and its germ could already be found in Friedrich Ratzel's (1848-1904) *Politische Geographie* (1897), but the subject is inextricably linked to the geopolitical thinker Karl Haushofer (1869-1946). Born in Munich, Haushofer spent several years in Japan while in the military. Japan's social structures, in which family, village community and state shaped the life of the individual, made a deep impression on him. He was actively involved in the creation of the *Zeitschrift fur Geopolik* (1924), and the *Akademie zur wissenschaftliche Erforschung und der Pflege des Deutschtums* (1925), becoming its president in 1933. He had previously befriended Rudolf Hess, an important friendship as it turned out, all the more so because Haushofer's father-in-law was Jewish. Hess's flight to England in May 1941 did create problems for Haushofer, but then Himmler took him under his wing, and Haushofer continued to place his geopolitics at the disposal of the Nazis. His oldest son, Albrecht (1903-45), an equally loyal devotee of the Third *Reich*, later became indirectly involved in the attempt on Hitler's life in July 1944. He was arrested, as were his brother and father, and shot dead just before the liberation. Haushofer himself was scheduled to be tried at Nuremberg as a war criminal, but the allies soon dropped the case against him. In the USA the idea had gained currency that before and during the war Haushofer had been in charge of a huge institute with thousands of staff that did nothing other than help Germany prepare for war. That was clearly not true, but the stake of geopolitics in support of Germany's struggle for *Lebensraum* should not be underestimated.

The publicist Rudolf Walther, in a survey article in *Die Zeit* of 21 July 1995 with the revealing title, "Man braucht mehr Platz", showed that geopolitics has again begun to rear its head, not just in Germany but also in France and Italy. Many nationalist and conservative scholars are attempting to strip geopolitics of its dubious past, or simply ignore it altogether.

15 Bakker, *Duitse geopolitiek*, p. 157.

Ostforschung

Ostforschung went beyond geopolitics in several respects. While the latter was primarily preoccupied with space, the former also took an interest in the people who lived there, identifying and classifying the population groups on Germany's periphery. *Deutschtumskunde* and Folklore had existed since before the establishment of the *Kaiserreich* in 1871, but now could be pursued in a broader context and be given more of a practical application. According to geopolitics, borders were not etched in stone, and a great number of German speakers were still living outside the *Kaiserreich*. Because ethnic Germans had to be protected from hostile majorities, as in Poland, Czechoslovakia and other border regions, *Volkstumpolitik* could operate at two levels: either importation, "*Heim ins Reich*", where minorities at the periphery faced extinction, or incorporation into the *Reich* of the areas containing ethnic Germans, as was done, for example, in 1938 with the Sudetenland. But in the *Kaiserreich* things had not yet reached this point. Through *Kolonialforschung* Germany tried to keep up with great powers like England and France, and with smaller countries like the Netherlands and Belgium. But as we have already noted, she had simply come too late to the table. The gaze had to be directed eastward; the East also contained more ethnic Germans than the West, so the *Neugestaltung* would have to take form in the East. But along which lines? Certainly, nothing along the lines of the Russian Revolution, despite its indirect contribution to Germany's remarkable eastward expansion towards the end of the war. The war had been lost, and with it enormous territories, including the few colonies in her possession. Scholars associated with the *Nordostdeutsche Forschungsgemeinschaft*, the *Deutsch-Akademischen Gildenschaft*, the *Verein für das Deutschtum im Ausland*, and the *Leipziger Stiftung für Deutsche Volks- und Kulturbodenforschung* were now able to combine their contempt of the Slav with a disgust for Communism. Historians, geographers, lawyers, pedagogues and geo-politicians spoke a similar language: *Lebensraum im Osten*. *Volk* and race increasingly took centre stage. From 1933 on, they were required to work closely with state and party organisations like the *Ostministerium*, headed by the Nazi ideologue and Baltic German Alfred Rosenberg (1893-1946), the *Reichskommissar für die Festigung deutschen Volkstums*, and the *Volksdeutsche Mittelstelle*. Not that these always worked together seamlessly; the SS and the party, for example, vied for control over *Volkstum*. What was at stake was nothing less than the future of Germany, the Thousand-Year *Reich*, of which the SS considered itself the visionary architect.

But the cohorts in the *Kämpfende Wissenschaft* camp could agree on one thing: there would be no place in the future German territories for "inferior" people. The Third *Reich* would be a racially pure Teutonic *Reich*, and veiled terms such as *Flurbereinigung*, *Aussiedlung* and *Entjudung* became the overtures to National Socialist policies in the East. Some scholars left their comfortable institutes during the war in order to ply their trade in the East. There they were able to satisfy themselves that in some places Jews had completely disappeared or were only to be found in ghettos. The classification of the population in some instances had proceeded to the point of dividing ethnic Germans into "still useful" and "not useful" categories. This latter group was then labelled "inferior," sealing its fate: "Vernichtung durch Arbeit".

Generalplan Ost was drawn up in 1941-42 and served as a blueprint for colonising and administering the occupied territories. The *Reichssicherheitshauptamt (RSHA)*, the security services headed by Himmler, had devised the plan, based on pre-war research into *Ostforscher*. There was a short-term and a long-term plan, with the latter scheduled to go into effect only after the war had been won. Its crux was the Germanisation of Poland. Poles with German ancestors would be given the "choice" of registering as ethnic Germans. Polish youngsters who looked sufficiently "German" were brutally separated from their parents, to be raised in Germany as real Germans. We will never know their number. Thousands of Polish intellectuals were murdered or put in camps. The remaining Poles would work as slaves for the Germans, or at least such was the plan. In the *Generalgouvernement* Polish farmers were driven from their land, after which the land was to be settled by German farmers. The *SS-Rasse- und Siedlungshauptamt* was in direct charge but competed, as a matter of course, with other party offices such as the *Deutsche Arbeitsfront* (*DAF*). It was not quite the hoped-for National Socialist Utopia they had had in mind, and we may safely speak of an "Entfesselte Technokratie."[16]

The same scenario was planned for White Russia and the Ukraine. There remained, naturally, many questions to be answered, one being the extent to which the remaining Slavs would have to be educated. "To count to ten and to obey their German masters", went the most extreme answer. But more sober minds realised that education was necessary in a modern society, even for simple activities. Theory and practice did not always mesh, and the conquered territories were enormously attractive to the petty power-mongers who loved to play lord and master and who took little notice of Berlin's directives.

[16] See Rolf-Dieter Müller, review of Machtild Rössler & Sabine Schleiermacher (eds.), *Der "Generalplan Ost" Hauptlinien der nationalsozialistischen Planungs- und Vernichtungspolitik*, Berlin, 1993, in *Die Zeit*, 25 March 1994.

Generalplan or no, the war had to be won as well. The optimism that marked the first months of Operation Barbarossa had long since evaporated, and of all the plans the only one still being pursued with alacrity was the annihilation of the Jews, gypsies and political commissars of the Red Army.

Westforschung

Westforschung seemed less harmful than *Ostforschung*, if only because no politics of destruction appeared to be involved. Yet the destruction of the Jews in Western Europe was entirely premeditated, and it was carried out.

Revising the western border was a principal preoccupation of the scholars of the *Bonner Institut fur die geschichliche Landeskunde der Rheinland* and the *Deutsch-Niederländische Institut* in Cologne, to mention but just a couple. Proof had to be supplied that Alsace-Lorraine, Eupany-Malmedy and the Saarland had always been German and thus had to come "*Heim ins Reich*". Surprisingly, the scholars also tried to show that Wallonia and northern France should be part of the Great German *Reich* as well. There was much more to it than revenge for "Versailles". France stood for revolution and, later, parliamentary democracy; Germanness wanted nothing to do with that. The battle with France (and with England as well) was also a battle for *Kultur: Kulturraum, Rasse* and *Volk* were important components of *Westforschung*, with the Dutch and the Flemish seemingly viewed as one step above other peoples. The special position of the Netherlands was very much in evidence when, six weeks after the attack in May 1940, the *Fuehrer* sent the Dutch prisoners of war back home. That brothers should be fighting one another was incongruous enough, but to keep them imprisoned for a long time was even more so. The *Fuehrer*'s magnanimity did not extend to the poor Flemish, however. In his speeches the *Reich* Commissar in the Netherlands, Dr A. Seyss-Inquart (1892-1946) continuously referred to ethnic affinities, and he seemed downright disappointed when Dutchmen did not share his views. Moreover, Germans frequently harboured quasi-positive feelings towards Dutch resistance fighters – as long as they were not Jewish or Communist – because they were considered to be "true" Germans. In 1942, Himmler himself expressed such sentiments in the course of a conversation with a resistance fighter. "You are Teutons, aren't you? So why are you fighting us?".[17] But this in no way lessened the punishment meted out to resisters.

---

[17] "Confrontation with resistance fighter Pim Boellaard", 14 May 1942, in J. van Bolhuis, *Onderdrukking en verzet: Nederland in oorlogstijd*, Vol. I, Arnhem, 1949, p. 598.

The German historian Franz Petri (1903-93) played a pivotal role. Petri earned his doctorate with a dissertation on the formation of language. His works include titles like *Germanisches Volkserbe in Wallonien und Nordfrankreich: die fränkische Landnahme in Frankreich und den Niederlanden und die Bildung der westlichen Sprachgrenze*,[18] and decades later, *Die fränkische Landnahme und die Entstehung der germanisch-romanischen Sprachgrenze in der interdisziplinären Diskussion*.[19] In 1938, he wrote that two things were important: research into the origins of Germanic-Romance linguistic separation and the Germanic *Grenzlandschicksal*, and the process whereby the Netherlands became an independent Germanic region.[20] Protestant by birth, he took his time embracing National Socialism, only becoming a true believer as the Nazis powered their way to success after success, taking out party membership in 1937. Just before the German campaign in May 1940, he wrote *Die Niederlände (Holland und Belgien) und das Reich*.[21] This booklet, a spare sixty pages, was informational material for the German military. Shortly before the war Petri became director of the *Deutsch-Niederländisch Institut*, and in the war itself he secured an important post with Von Falkenhausen's *Militärverwaltungsstab* in Brussels. He now was a convinced National Socialist of the SS orientation, with emphasis on the *Völkische*, and it is curious how relatively mildly people judged him later on.[22] After the war Petri, with some assistance from his Belgian and Dutch colleagues, was quickly rehabilitated and became a professor in Bonn in 2001. His post-war writings differ but little, except in tone, from his work before 1945, but his case, however, is by no means unique in post-war Germany.

A conspicuous outsider in the historiography of *Westforschung* is Hans Derks, whose study *Deutsche Westforschung* was published in 1961.[23] In addition to an extensive overview, the work also contains a fierce attack on a number of Dutch scholars who agreed with Petri on a

---

18 2 Vols, Bonn, 1937 & 1942.

19 Darmstadt, 1977.

20 I. Schöffer, *Het nationaal-socialistische beeld van de geschiedenis der Nederlanden*, Arnhem/Amsterdam [1956], p. 103.

21 *Die Niederlände (Holland und Belgien) und das Reich. Volkstum, Geschichte und Gegenwart*, Bonn, 1940.

22 Schöffer, *Het nationaal-socialistische beeld*, p. 327. See, for example, Karl Ditt, "Franz Petri und die Geschichte der Niederlande. Vom germanischen Kulturraum zur Nation Europas", *Tijdschrift voor Geschiedenis*, 118/2 (2005), pp. 169-187, where he replaces *Westforschung* with the more innocent-sounding term *Kulturraumforschung*.

23 H. Derks, *Deutsche Westforschung. Ideologie und Praxis im 20. Jahrhundert*, Leipzig, 2001.

number of points, both immediately after the war, and up to the present day. He oversteps the mark in characterising the German institutes founded in the Netherlands in the 1990s and the *Zentrum für Niederlände-Studien* in Munster as Petri's direct successors, and thus collaboration after the fact. Those institutes have never indulged in imperialistic projects or expressions of *Grossraum*. For these reasons Derks' book has gone practically unnoticed, with the exception of some negative reviews.[24]

In a physically monumental collection of 2003, *Griff nach dem Westen*, more than forty authors treat numerous aspects of *Westforschung.*[25] As the title indicates, it deals not just with scholarship but also with German expansionism. The emphasis is on the Netherlands and Flanders, and the number of Dutch authors is considerable. The absence of France is unfortunate, because that is precisely where the problem resided. For the Netherlands and Flanders one might advance a claim of "ethnic kinship", but to make Latin France an essential part of Germany is far more challenging. The Netherlands and Flanders were viewed positively, while in these terms France was not; nonetheless all three would have to be incorporated into the Third *Reich* somehow. No such luck for poor Luxembourg: she was given short shrift during the war and simply incorporated into Germany.

The Netherlands: a Digression

It should not be forgotten that, before the war, Dutch universities and Dutch scholars looked towards Germany. This was true of textbooks, educational exchanges, obtaining a PhD from prestigious German universities, and so forth. On an intellectual level, the Dutch and the Germans knew each other well. This did not end overnight with Hitler's accession to power in 1933.

The eleventh War Documentation Yearbook of 2000 was devoted to national characteristics, race, culture and science in the Netherlands in the first half of the twentieth century.[26] It unambiguously posited the connection between scholarship and politics. Proceeding from different

24 The very critical review by Gjalt Zondergeld in *Bijdragen en Mededelingen betreffende de Geschiedenis der Nederlanden*, 119/4, 2003, pp. 598-601, was trumped by Klaas van Berkel, who condemned the book as "slander", in *Tijdschrift voor Geschiedenis*, 108/2, 2005, p. 273.

25 Burkhard Dietz *et al.* (eds.), *Griff nach dem Westen. Die "Westforschung" der völkisch-nationalen Wissenschaften zum nordwesteuropäischen Raum (1919-1960)*, 2 Vols, Munster, Waxmann, 2003.

26 *Volkseigen, ras, cultuur en wetenschap in Nederland 1900-1950; Elfde Jaarboek Oorlogsdocumentatie*, Zutphen, 2000.

disciplines such as history and anthropology, the discourse covered a broad spectrum: archaeology and propaganda, folklore, the racial discourse within the social sciences and racial hygiene. The contributors sought to find out whether "in the Netherlands, too, racial thinking of the National Socialism stripe and certain variations of colonialism were not more deeply embedded in scholarly thinking than was later assumed".[27]

Ostforschung continued to be carried out during the war, albeit on a small scale. Its primary goal was to motivate the Dutch to volunteer to serve in the German-occupied eastern territories. The Dutch, after all, had been there before as Germanic pioneers, on the banks of the Weichsel (Vistula), in Poland and in Brandenburg;[28] indeed, the sentiment found expression in Dutch doggerel:

> Naar Oostland willen wij rijden
> Naar Oostland willen wij mee
> Al over die groene heiden
> Daar isser een betere stee
> [To Eastland we want to ride
> To Eastland we want to go
> Across heaths of green
> For better farmland, that's our dream].

The renewed Dutch interest in the East was based on two factors: the German need for plentiful capital and Germanic workers in the occupied lands in Eastern Europe, and the loss of Dutch colonies to Japan. Moreover, M.M. Rost van Tonningen (1894-1945), the third man in the National Socialist movement in the Netherlands, did not believe a Teuton's place was in the tropics, even though he himself was born in the Dutch East Indies, as the youngest son of the supreme commander of the Dutch colonial army (KNIL). His gaze, too, was fixed on Eastern Europe. The interest in his *Nederlands Oostcompagnie* (Netherlands Company of the East) would be fostered by "historians" like Johannes Theunisz, an SS man who contributed a number of works on the history of the Dutch in Eastern Europe, including a *Bibliographie van Belgische en Nederlandsche geschriften betreffende vestiging van Nederlanders in*

---

27 *Ibid.*, p. 8.

28 See David Barnouw, *Oostboeren, zee-germanen en turfstekers*, Amsterdam, 2004. For a broader context, see Rolf-Dieter Müller (ed.), *Die deutsche Wirtschaftspolitik in den besetzten sowjetischen Gebieten 1941-1943. Der Abschlussbericht des Wirtschaftstabes Ost und Aufzeichnungen eines Angehörigen des Wirtschaftskommandos Kiew*, Boppard am Rhein, 1991.

Midden- en Oost-Europa,[29] and *Naar Oostland. Nederlandse kolonisatie in Europa.*[30] Commissioned by Seyss-Inquart himself, a collection of some 400 pages called *Die Niederlande in Umbruch der Zeiten*[31] appeared halfway through 1941. It featured a short article by H.W. van Etten (1916-2004)[32] entitled "Naar Oostland", and it contained the following ringing appeal:

> Once again the Netherlands is experiencing a shortage of land, and once again the call comes from the East for vigorous young men of a farming bent who are prepared to resume the tradition of their forefathers and to fight for their future in a new land.[33]

Colonisation of the East did indeed attract some interest from the Netherlands. There were farmers who went to harvest "grain for the Netherlands", followed by fishermen, peat cutters and construction workers. These went to try their luck in the Baltic countries, White Russia and the Ukraine, enticed by promises of a glorious future. There were National Socialists among them, but just as many adventurers and black marketeers, as well as Dutchmen who tried to avoid mandatory labour service in Germany. There was no dearth of plans: the transfer of Dutch sugar refineries to the East; the siting of vast peat farms; the "purchase" of local brick factories and drainage works on the Dnieper. The *Oostinstituut* in The Hague was to pave the way for all these plans and furnish the requisite ideological background. In reality, conditions in the East were altogether different. The locals were not the only ones who did not care for the Dutch interlopers; their fellow Teutons from Germany looked down on the Dutch as well: defeated in five days in May 1940 and now they were to be allowed to participate as equals in the East? The Dutch were quick to realise that there would be no equal treatment, and that the promised farm with a docile workforce was a mirage. Black-market activities eventually became so widespread that the Germans dubbed the Netherlanders "white Jews". The Dutch suffered death at the hands of the partisans, and the Red Army was back far sooner than expected; the entire enterprise was a complete fiasco.

Naturally, *Westforschung* was far less evolved than *Ostforschung*, if only because the ultimate consequence would be the dissolution of the Dutch state. Rost van Tonningen was very much in favour, and in

---

29 Amsterdam, 1941.

30 Amsterdam, 1943.

31 M.F. du Prel & W. Janke (eds.), *Die Niederlande in Umbruch der Zeiten: Alte und Neue Beziehungen zum Reich*, Würzburg, 1941.

32 He was identified simply as *Schriftsteller*, but he was also involved with the Dutch SS and later the *Waffen-SS*.

33 Du Prel & Janke (eds.), *Die Niederlande in Umbruch der Zeiten*, p. 184.

speeches to NSB audiences repeatedly reminded them that the Netherlands would become an inseparable part of the Great German *Reich*. He continually made the point that the Netherlands, especially after the Peace of Munster in 1648, which brought to a close the Eighty Years' War with Spain, had strayed from her German brother, increasingly becoming a vassal state of England and France. The fact that the NSB Party Leader Anton Mussert (1894-1946) disagreed, and favoured a Germanic federation of states, did little to enhance Rost's standing with his NSB superior. Nonetheless, Rost van Tonningen wrote to Mussert in December 1940 that "The growing Jewish hegemony and the treason of the Netherlands' capitalist bourgeoisie" had been nipped in the bud by the "rebirth of National Socialism as the original form of liberation of the individual of Germanic blood".[34]

At end of 1942, the University of Utrecht's Faculty of Law (originally Leiden was to have enjoyed this privilege) appointed a lecturer in geopolitics, Dr J.G. Loohuis. He had earned his doctorate in 1931 with a thesis on South-East Asia,[35] and his first public lecture bore the title, "About Political Ignorance and the Science of Politics".[36] The Netherlands Institute for War Documentation in Amsterdam has a number of pamphlets by this formerly pro-German editor-in-chief of the *Utrechts Dagblad*, among them many reprints from the equally pro-Nazi *Haagsch Maandblad*. Dilettantes who dared attack the "real" geopolitics were favourite targets. His writings also paid extensive tribute to the Swede Rudolf Kjellén, "the architect of the great power studies". In the course of his trial in 1947, Loohuis admitted having joined the *Nederlands-Duitsche Cultuurgemeenschap* after attending a lecture by Haushofer. The tribunal found that he had obtained his lectureship "by publicly currying favour with the enemy". Moreover, by contributing articles to collaborationist newspapers, he had demonstrated a "worryingly un-Dutch mentality". Loohuis was ultimately incarcerated for three years, and deprived of the right to vote and the right to work in the public sector.[37] Despite his antics, however, a study by Kees Bosman has shown that geopolitics gained but few adherents in the Netherlands.[38] Jillard Keijzer ascribes this – in part – to the fact the Netherlands was a colonial power. The German invasion and the loss of

34 Rost to Mussert, 5 Dec. 1940, in E. Fraenkel-Verkade (ed.), *Correspondentie Mr. M.M. Rost van Tonningen*, Vol. I, The Hague, 1967, pp. 535-536.

35 *Landbouwgordel en de industriecentra der gematigde zone in hare beteekenis voor het huidige aspect van het Koloniale vraagstuk in Zuid-Oost-Azië*, Hilversum, 1931.

36 *Utrechtsch Nieuwsblad*, 27 Nov. 1942.

37 *Nieuw Utrechts Dagblad*, 12 Sept. 1947.

38 *Ruimte voor een nieuwe tijd: vormgeving van de Nederlandse regio 1900-1945*, Rotterdam, 1993, especially pp. 293-299.

the East Indian colonies did, however, lead to an increased Dutch interest.[39] From the twenty-first century the Nazi dreams about a New Europe can be seen as marginal, because Germany lost the war and that was the price to be paid by the loser. But those apparently wild ideas were evolved in the largest and – arguably – the most highly developed nation in Europe at that time.

The concepts of *Ostforschung* and *Westforschung* can no longer be used without a historical explanation (and apology). Nevertheless, the taint associated with geopolitics has all but vanished. Anti-globalists often use the term, and it routinely surfaces in articles about the Russian Federation in the Dutch liberal daily the *NRC/Handelsblad.* In 2005 the University of Groningen appointed a Professor of Geopolitics and Energy Management funded by the Netherlands Gasunie, without a single individual dredging up Haushofer or the Second World War. Today geopolitics is ever more associated with raw materials and their distribution on a global scale. We owe to the late beer baron F. Heineken (1923-2002) an innocent post-war variation on geopolitics. In 1992, worried about the fate of Europe – Yugoslavia was about to fall apart – he conceived *Eurotopia*, a new map of Europe consisting of seventy-five new countries, each with five to ten million inhabitants. He desired stronger "self-identification", increased security, and more efficient government. On the whole, these projected "new" countries coincided with existing regions, regions in which "self-identification" usually poses few problems.

We can conclude that geopolitics is experiencing a certain revival, but one in which the concept of *Lebensraum* no longer plays a role: there may be a role for it in the settlement of the affairs of the former Yugoslavia. To the extent that research is still being done in the field of *Ostforschung* and *Westforschung*, it is of a purely academic character, without expansionist goals or border revisions at the expense of neighbouring countries. Borders are not immutable: the large-scale border changes after the First and Second Wolrd War prove as much, although until the crisis of Yugoslavia it seemed that we had reached a status quo of long duration. Historians cannot and should not predict the future, but it would appear that "classical" *Ost-* and *Westforschung* are now being invoked primarily in the interests of coming to terms with the past.

---

39 "Haushofer in Holland. Een studie naar de geopolitieke visie binnen de NSB", unpublished *doctoraalscriptie* [MA thesis] Political Geography, University of Amsterdam, 2002.

Part III

From the "Margins" of Europe

CHAPTER 4

The Turkish Perception of Europe

Example and Enemy

Erik-Jan ZÜRCHER

To say that Turkey wants to be a part of Europe is to state the obvious. For years now, debates on the possible future accession of Turkey to the European Union have filled our newspapers. Many experts and politicians have pronounced on the political, economic, security-related and cultural arguments for or against accession. It is often pointed out in the newspaper articles, particularly those written by supporters of the Turkish case, that Turkey's membership ambitions have old roots. Turkey had already been a full member of the North Atlantic Treaty Organisation for seven years when, along with Greece, it applied for associate membership of the European Economic Community in 1959. Under the Treaty of Ankara of 1963 Turkey was granted associate status as a first step towards the full membership that was defined as the ultimate aim in the treaty. Europe accepted the argument that a country that was vital for the defence of the "Free West" should also be strengthened economically by integration with the West.

The first decade following the conclusion of the treaty can be characterised as a honeymoon period in the relationship between the European Community and Turkey, but after the almost simultaneous first oil crisis of 1973-74 and the Turkish invasion of Cyprus in 1974, the relationship entered a period of slow progress and frequent setbacks. Not only did the oil crisis hit the Turkish economy very hard (because it had to come up with four times the amount of hard currency to foot its oil bill), it also led almost overnight to the destruction of old, labour-intensive industries in Northern Europe (mines, shipyards, textile factories) that employed huge numbers of Turkish "guest workers". Before 1974, safeguarding a continued supply of migrant labour from Turkey had been one of the motives for European states to further Turkish integration into Europe. After 1974 the Turkish migrant workers, many of them now out of a job, came to be seen as a problem. At the same time, the Turkish invasion of Cyprus, and in particular its continued occupation of forty per cent of the island, isolated Turkey politically. The military *coup d'état* of September 1980 restored Turkey's stability and ultimately its economy as well, but it meant another setback for Turkish-European relations. Europe obviously could not negotiate about accession with a military dictatorship. The full effect of the "lost years" was felt when Turkey's new civilian Prime Minister Turgut Özal, the architect of Turkey's market economy, applied for full membership in 1987. Although the original Treaty of Ankara had foreseen the accession round this time, Turkey's bid was rejected on the basis of a negative opinion on the part of the Commission. After another rocky patch in the mid-1990s, when European politicians like Kohl, Bolkestein, Martens and Giscard d'Estaing began to debate openly whether Turkey could qualify as a European country for cultural reasons, the die was finally cast at the Helsinki summit of December 1999, when Turkey was accepted as a candidate for membership, to be judged by the same criteria that were applied to the other East-European and Mediterranean candidates. We all still remember that the European heads of government decided to start accession negotiations at the The Hague summit of 2004, and that negotiations were actually launched on 3 October 2005.

All the different steps towards integration with Europe have been celebrated like so many national successes in Turkey. When in 1995 the customs union between Turkey and the EU was signed, Turkish newspaper headlines shouted "Finally!" and "We have entered Europe" (although opposition leader Yılmaz made dark references to "Baltalimanı", the name of the free trade treaty of 1838 by which the Ottoman Empire had opened up its markets to British commerce and lost its monopolies). The Turkish elite in particular saw and sees the accession of Turkey to the EU as proof that the country has reached its destination; that it has become a secular, democratic and above all "modern" and "civilised" country. It is not hard to see why this should be so. In background articles in the European press the reader is quite often reminded of the fact that the Turkish ambition to be a part of Europe, or "the West" is not a new phenomenon; that it is in fact much older even than the start of the accession process in 1963.

From the start of the nineteenth century, the Ottoman governing elite, faced with a series of disastrous wars against the Russians and with the French occupation of the Ottoman province of Egypt, realised that, if the empire were to be saved, it had to adopt European technology, particularly military technology. The building of a modern Western-style army brought with it the necessity to create a new educational system, more efficient taxation, conscription and much else besides, all of it borrowed from Europe.[1] By the mid-nineteenth century European administrative techniques, but also political ideas like the social contract, constitutionalism and patriotism began to spread among the Ottoman elite.[2] By the turn of the twentieth century, decades before the proclamation of the Turkish Republic in 1923, much of the legislation was based on European positive law and was entirely secular. European bourgeois culture had entered the cities of the empire and places like Salonica, Izmir, Beirut and, of course the capital Istanbul had their tearooms, parks, gentlemen's clubs and sports clubs, but also trams and water companies.[3]

In the 1920s and 1930s of the last century this process of reform according to European models culminated in the "social and cultural revolution"[4] unleashed by the founder of the Republic of Turkey, former Ottoman brigadier Mustafa Kemal Pasha, in the fifteen years between 1923 and his death in 1938. The policies of Atatürk ("Father Turk", the name given to him by parliament in 1934) in part were a continuation of those of the late Ottoman era. He completed the Westernisation of the institutions and the legal system, with the formal abolition of the *Sharia* and its replacement with the Swiss civil code in 1926 as perhaps the most radical step. Atatürk aimed radically to transform both the world view and the way of life of the citizens of Turkey. That was an ambition he shared with many other "Young Turks" of his generation, but where nearly all of them had sought to achieve some kind of synthesis between "Eastern" and "Western" elements, Atatürk was more radical. He wanted to turn his compatriots into Europeans both inwardly and outwardly and while changing their world view was of necessity a long-term project, the attempt of the Kemalists (as Atatürk's followers were

1 See: Stanford Shaw, *Between Old and New. The Ottoman Empire under Sultan Selim III 1789-1807*, Cambridge Mass., Harvard UP, 1971; and Carter V. Findley, *Bureaucratic Reform in the Ottoman Empire: the Sublime Porte, 1789-1922*, Princeton, Princeton UP, 1980.

2 See: Şerif Mardin, *The Genesis of Young Ottoman Thought. A Study in the Modernization of Turkish Political Ideas*, Princeton, Princeton UP, 1962.

3 See: Fatma Müge Göçek, *Rise of the Bourgeoisie, Demise of Empire: Ottoman Westernization and Social Change*, Oxford, Oxford UP, 1996.

4 Atatürk himself often used the term "fikrî ve ictimaî inkılap", which he equated to "the march along the road to civilisation and renewal" (medeniyet ve teceddüd yolunda yürüme). See for instance his speech of 20 September 1924 in Samsun in *Atatürk'ün Söylev ve Demeçleri II (1906-1938)*, Ankara, TTK, 1959, p. 191. The translation of the word *inkılap* poses a problem as, like "revolution", it derives literally from the stem 'revolve', but carries the meaning of fast, radical but orderly change. Late Ottomans used a different word, *ihtilal*, for political revolutions like that in France. The question whether *inkılap* may be translated as revolution has been hotly debated between the right and the left in Turkey.

known) to transform the appearance of Turkey and its citizens was quite successful, having immediate effects, at least in the towns.[5] The introduction of European clothing, the European alphabet, clock, calendar, Sunday rest, weights and measures, but also the introduction of ballroom dancing, the shaving off of moustaches and beards all served the same purpose: to turn Turkey into a "civilised" country. I use the term on purpose, because the Kemalists were motivated first and foremost by a civilisational ideal, and if there is anything that sets Atatürk and his followers apart from their immediate Ottoman predecessors, it is their belief that there was but one single viable civilisation in the world: that of Europe. If Turkey was to survive at all, it could only do so by becoming European lock, stock and barrel, and not through the creation of hybrid forms of civilisation in which East and West were somehow combined. Atatürk was untiring in his efforts to spread this message and the words "modern" (*asri, muasir*) and "civilised" (*medeni*) crop up constantly in the speeches he held all over the country in the 1920s and 1930s, and their meaning is more or less interchangeable. A prime example of this usage is the speech he held on 29 October 1933 at the celebrations of the tenth anniversary of the republic. This speech is quite famous in Turkey and its final sentence, "Happy is the one who can say, 'I am a Turk'" (*Ne mutlu Türk'üm diyene*) is a catchphrase that adorns countless monuments, government buildings and even whole mountainsides.[6]

In this ten-minute speech, delivered from the rostrum in a packed sports stadium in Ankara, the terms "civilisation" and "civilised" occur no less than eight times. It would go too far here to translate the whole text, but two samples suffice to illustrate both the content and the tone. In the first part of his speech (not quoted here) Atatürk refers to the struggle for independence after World War I and to the successes already achieved, but then he goes on to say that that is not enough. It is only a beginning and the Turks must work hard to achieve even greater things in the future. What they are to achieve is nothing less than to become civilised: "We shall elevate our national culture onto the level of contemporary civilisation", and:

---

5 Erik-Jan Zürcher, "Two Young Ottomanists Discover Kemalist Turkey. The Travel Diaries of Robert Anhegger and Andreas Tietze", *Journal of Turkish Studies*, 26/1, 2002, pp. 359-369.

6 The Tenth Anniversary Speech (Onuncu Yıl Nutku) has been reproduced in many places and it is also one of the very rare speeches by Atataürk of which a sound recording exists. I have used the text reproduced in the album put out by the party on the occasion of the fifteenth anniversary of the republic in October 1938. Atatürk was by then very ill (he would die a fortnight later) and could not deliver a speech, so his speech of five years earlier was used. See Cumhuriyet Halk Partisi, *On Beşinci Yıl Kitabı*, Ankara, 1938.

> Today I say with the same conviction and the same certainty that the whole civilised world will soon recognise once again that the Turkish nation, which marches fully united towards the national ideal, is indeed a great nation. I don't doubt for a minute that the forgotten, but great civilising force and the vast civilising ability of the Turkish people, once discovered, will rise like a new sun on the horizon of the exalted civilisation of the future.[7]

Of course, the rhetoric sounds overblown to twenty-first century ears, but it dates from 1933, and here we have a head of state, a military officer who is to all intents and purposes a dictator at the height of his power; a contemporary of Hitler, Mussolini and many other European dictators of the interbellum. He is celebrating ten years of successful rule and what does he talk about? Hardly at all about past victories, not about territorial ambitions or struggle against internal and external enemies, but about his confidence that the Turkish nation will become civilised. So civilised, in fact, that the "whole civilised world" will recognise it as a new force for good. In other words: the goal this national leader sets his people is nothing less than to become European with the Europeans.

Atatürk died in 1938, but he is still omnipresent in Turkey today. His effigy is everywhere and everyone taking part in political debates sooner or later refers to Atatürk's ideas as inspiration and guideline for the future. Small wonder, then, that in Western eyes, Turkey has the image of a country that aims to become European almost at any price and that the European political elites with European public opinion in their wake thinks that Turkey stands waiting at the doorstep, cap in hand. Although the arguments are cast in slightly different terms – such as the compatibility, or otherwise, of Islam and democracy, the Turkish record on human rights or the treatment of women – ultimately the Europeans also see it as a civilisational issue, and assume that it is up to Europe to determine whether and when the Turks have proved to be civilised enough to become part of the European family of nations. This, however, is at best only part of the picture. If our European newspaper readers were to follow the Turkish dailies for a while, they would be in for a surprise. Reading the editorials and the columns (of which several dozen appear in the Turkish press every day) would be especially shocking. What they would see is that the reporting, but especially the editorial comments and the analyses of the column writers, constantly refer to the way Turkey is under threat from the West – the USA but also Europe. Writers constantly refer to schemes and plots that are being hatched in order to weaken and ultimately perhaps even split up the country. When the US Congress or the French Parliament threatens to

7 *Ibid.*

pass motions on the Armenian genocide of 1915, it is seen in this light. When EU delegations investigate the human rights situation in Turkey, in particular with regard to the rights of linguistic or religious minorities, the newspapers also often see it as attempts by Europe to find pretexts to interfere and to undermine the "unity and indivisibility of the nation".[8] UEFA referees who supervise matches of Turkish clubs in the European competitions are regularly accused of possessing a "crusader mentality". "Valley of the Wolves", a Turkish film and television series with a fiercely nationalist and anti-Western message, is built around an incident during the Iraq War when American troops arrested and (in Turkish eyes) humiliated a group of Turkish officers in Kirkuk, and was a box office hit in 2005-06. "Those Crazy Turks", a romanticised version of the Turkish independence war against the Greek occupation in the years after World War I has been topping the bestseller lists for months. Anyone who follows the Turkish media for any length of time and takes note of productions like "Valley of the Wolves" must come to the conclusion that the events reported are placed in a historical context that makes them part of a "grand narrative" in which the West dislikes Turkey, and ultimately aims to keep the country small and weak.

The anti-Western undercurrent in the Turkish media and in Turkish public opinion is partly conjunctural. It is caused by the deeply unpopular American-led invasion of Iraq and by what is seen as the continued rejection of Turkey's European credentials by European politicians like Kohl, Giscard d'Estaing, Sarkozy, Bolkestein, Martens and Haider in the last ten years. The fact that one of the outcomes of the war in Iraq has been close co-operation between the US government and the autonomous Kurdish region in Northern Iraq has aroused the suspicion that the US is ready to bet on the Kurds rather than on Turkey as its main ally in the Muslim Middle East. So current events are partly responsible for the surge in anti-Western sentiment in Turkey, but the widespread nature and persistence of this perception indicates that it feeds on a much deeper, historically rooted image of the West as a potential threat, and it is these historical roots that I now want to investigate.

Earlier we saw how Atatürk's 1933 speech symbolised the ambition of the Turkish elite to become part of European civilisation. It is not difficult, however, to come up with an equally important historical text that represents the image of the West as a threat and rejects European

---

8 *Millî birlik ve beraberlik* is a term used often in Kemalist discourse, in particular by the army. Politics on the basis of ethnicity, religion or class are seen as contrary to this principle of national unity and therefore unlawful.

civilisation. This text, perhaps surprisingly, is that of the national anthem of the Republic of Turkey.

The Turkish national anthem, the "Independence March" (*İstiklal Marşı*) dates from the winter of 1921, or, to be more precise, its text does, as the melody was decided upon much later. It was a time of crisis for the Turkish nationalists. The west of the country was occupied by Greek forces, who were expected to advance further inland, as indeed they would do in the summer of 1921, when they came to within 70 kilometres of Ankara before being beaten off. In this tense situation the Turkish national assembly felt the need for a national anthem to inspire the troops and to mobilise the population.[9] A competition was launched, with a substantial prize. Over 700 poems were sent in, but one of Turkey's best-known poets, Mehmet Akif, did not join the competition even though he was a member of the assembly himself, because he felt that it was unseemly to gain a financial reward for what should be a patriotic labour of love. When pressed by the Minister of Education, Akif finally agreed to submit a text. This text was read to the assembly on 12 March 1921 and immediately received a rapturous welcome. The declamation was repeated several times to applause and tears from the audience. Akif was declared the winner of the competition, but true to his principles, refused the prize.

The Independence March has remained Turkey's national anthem ever since, even though its author, a devout Muslim, felt compelled to leave the country once Atatürk started his secularisation drive. He settled in Cairo and returned only shortly before his death in 1936. The text of the anthem consists of ten stanzas and is written in classical Ottoman verse. According to the standard protocol for official occasions in Turkey, only the first two stanzas are sung. They convey a message of endurance and patriotism. The Turkish nation will fight to the last man because independence is its right. But what do we see, when we read on, for instance in the fourth stanza?

> Even if the western horizon is surrounded by a wall of armour plate,
> My heart filled with belief forms a defensive line.
> You are strong, do not fear! How can such a faith be strangled
> By the toothless monster that you call 'civilisation'?[10]

---

9 The late Ottoman Empire did adopt the paraphernalia of the nineteenth-century state, with flags, emblems, uniforms and decorations. It also used marches composed first by European musicians like Gabriele Donizetti, brother of the better-known Gaetano, and later by Ottoman composers, but an official national anthem was never adopted.

10 This translation is of course a fairly literal prose rendering of the text. To get an idea of the original poem, I add the Turkish text of stanza 4:

The civilisation marching in from the West, extolled so loudly by Atatürk twelve years later, is here represented as a toothless monster that must be fought (and will be defeated) with the weapons of the faith. Atatürk's exalted civilisation, the West, is in Akif's view simultaneously an aggressive force that has invaded the fatherland and a sham that is no match for the religious fervour of the Turks. This enormous difference can be attributed to the different circumstances in which the two texts came into being. In 1933 a decade had gone by since the end of the independence war. Turkey's independence had been saved and the leadership was free to concentrate on what Atatürk called the "social and cultural revolution", the modernisation and secularisation of Turkish society. In 1921 the future was completely uncertain and it was an open question whether an independent Turkey had any future at all. The capital, Istanbul, had been occupied by the British a year earlier and the Greeks, who had landed in Izmir in 1919, had gradually occupied the most developed areas of the country. French troops still occupied parts of the south. The nationalists, with their back to the wall, did what they could to mobilise the Muslim population of Anatolia. As the concept of "nation" or "national self-determination" held no meaning for the largely illiterate peasant population, the mobilisation took place on the basis of religion, by referring to the struggle as one between Islam and infidel.[11]

The aim of the struggle was to reverse the death verdict that had been pronounced on the Ottoman state. Less than a year earlier, in July 1920, the victors of World War I had forced the sultan's government to sign the Treaty of Sèvres. Under this treaty the Ottoman Empire was divided up among the victors themselves and their clients. The French received a mandate in Syria and Lebanon and a zone of influence in Southern Anatolia; the British were given Iraq, Jordan, and Palestine. A large independent Armenia was created in Eastern Anatolia, the Kurds were given autonomy with the option of independence (except, of course, in British Iraq), Italy received a large zone of influence in the south-west (based on the port of Antalya), Western Anatolia as well as European Turkey was awarded to Greece, and the straits were internationalised. All that remained for the Turks was part of Central Anatolia.

Garbın âfâkını sarmışsa çelik zırhlı duvar.
Benim iman dolu göğsüm gibi serhaddim var.
Ulusun, korkma! Nasıl böyle bir imânı boğar,
"Medeniyyet!" dediğin tek dişi kalmış canavar?

11 Erik-Jan Zürcher, "Young Turks, Ottoman Muslims and Turkish Nationalists: Identity Politics 1908-1938", in Kemal H. Karpat (ed.), *Ottoman Past and Today's Turkey*, Leiden, Brill, 2000, pp. 150-179.

Of all the peace treaties concluded after World War I, Sèvres was by far the harshest.

For the Ottoman Turks, Sèvres was not only a disastrous conclusion to a war, but the final act in a tragic sequence of events. In ten years the Ottomans had lost three wars in quick succession. In 1911 Italy, on the flimsiest of pretexts, had invaded the last North African province of the Empire, Tripolitania (modern Libya), initiating a war between the two countries. That war was still being waged when four Christian Balkan states, Greece, Bulgaria, Serbia and Montenegro, jointly attacked the Empire, routed its army and divested it of its last European provinces. The areas that were then lost had often been in Ottoman hands for five hundred years, and a large part of the Ottoman elite hailed from these European possessions. The Balkan War had been over for slightly over a year when World War I broke out. Small wonder, then, that Sèvres (*Sevr* in Turkish) has become a code word, a deeply emotive term that signifies loss and defeat, and refers to a period in which the country was forced time and again to accept the loss of territories under humiliating treaties, caused by European aggression, often in combination with secessionist movements among the Empire's minorities.

This memory lives on. Undoubtedly, Atatürk's ambition to turn Turkey into a strong, modern country after West-European examples, has taken root there. In Atatürk's lifetime and under his successor İsmet İnönü, his brand of modernisation may still have been a recipe prescribed by a small urban elite to the population at large, but over the past fifty years it has gradually been appropriated by the growing middle class that has a different interpretation of the concept of secularism, but an equally ardent desire to become part of Europe. But the other side of the coin, the deep-rooted suspicion of Europe that is symbolised by the name of Sèvres, is also much in evidence today. Former president Süleyman Demirel more than once referred to what he called "the new Sèvres" (*yeni Sevr*). The context in which he used the term was that of the Kurdish problem, more particularly the defence of Kurdish minority rights by the European Parliament and several European governments at a time when Turkey was fighting the armed insurrection of the Kurdish separatist movement PKK. He saw clear parallels between this situation and the European espousal of the Greek and Armenian causes in 1920. Over the last decade the term *Yeni Sevr* has gained wide currency among nationalists in Turkey. Ironically, these are not found in the ranks of the governing AK party with its Islamist roots, but in those of the parties that regard themselves most explicitly as the guardians of Atatürk's legacy, the ultranationalist MHP and the Republican People's Party (CHP), founded by Atatürk himself. An internet search quickly shows how widely the term is used, and it also makes clear that for some

Turkish nationalists it is no longer just the Kurdish question, but the very process of accession to the EU itself that is seen as a European attempt to achieve what it had tried but failed to gain in 1920.

For those who are directly involved in the process of negotiations about Turkish accession to the EU, but also for those who want to understand what is involved in this process, it is important to realise that Turkey's picture of the Europeans is the complicated result of conflicting historical legacies, deeply rooted in the collective memory of the Turkish people. The Cyprus issue is a case in point. Turkey has often been blamed for its intransigence in the protracted negotiations on the future of the island, and the refusal to open up Turkish ports and airports to Cypriot shipping has already led to significant delays in the accession negotiations. Even European observers who know the complex history of the Cyprus problem, and who appreciate that the European Union has been far from blameless, are puzzled by the fact that Turkey seems to put into jeopardy its European destiny for a sensitive, but ultimately minor issue. They realise that for Turkey strategic considerations play a part, and that it feels a responsibility for the Turkish community in Northern Cyprus, but there are several German cities that have larger Turkish populations than the whole of Cyprus!

Looked at from the Turkish side, the picture is different. Cyprus had been Ottoman territory for three hundred years when the British forced the Sultan to cede it to them as the price for their intervention to undo some of the effects of the disastrous Ottoman-Russian war of 1877-78. With its back to the wall, the Ottoman government had had no choice but to give in. When Britain decided to end its rule on Cyprus in the late 1950s, the island was not returned to Turkey (as the successor to the Ottoman Empire), but granted independence under a government dominated by the Greek majority. No sooner had the island gained independence than its government set out to limit the constitutional guarantees of the Turkish minority. Muslim Turkish villages were attacked by armed gangs of EOKA, the Greek nationalist movement, and finally, in 1974, a *coup d'état* by Greek Cypriot army officers toppled the government and announced union with Greece. All of this for the Turks was a *déjà vu*. It reminded them of the way their grandparents had been terrorised and forced to flee when the Christian nation states in the Balkans came into being in the nineteenth and early twentieth centuries. For at least a third of the population of present-day Turkey, the exodus from the Balkans, the Black Sea littoral or the Caucasus is part of their own family history, something that helps to explain the strong bonds of solidarity felt with the pockets of Muslims that have survived the emergence of Christian nation states in the Balkans: the Bosnians, the Kosovars, the Turks and Pomaks of Bulgaria, the Turkish-speaking community in north-eastern

Greece. In Turkish eyes, the Turkish community in Cyprus falls into this category and it is a national duty to protect it. Turks are well aware that public opinion in EU member state Greece supported the policies of Slobodan Milosevic in Bosnia and Kosovo. This makes it all the more remarkable that Turkish Prime Minister Tayip Erdoğan and his foreign minister Abdullah Gül were ready to accept the Annan plan for reunification of Cyprus in the face of strong opposition from the Turkish army, Turkish nationalist leaders in Northern Cyprus and large sections of their own public opinion. Unfortunately the duplicity of the Greek Cypriot government scuppered the deal and the European Union, which had repeatedly stated that it would only accept a united Cyprus as a member, succumbed to Greek pressure and accepted the divided island.

In conclusion we can say that the prevailing image of Turkey as a Europhile country needs some correction. Certainly, there is a genuine desire to become part of the European Union because it is felt that it will make Turkey's development towards a strong, secular and modern state irreversible. It will set Turkey apart from the disaster areas to its south and east. On the other hand the collective memory of European support for secession of minorities from the Ottoman Empire and of European attempts to wipe the country off the map after World War I is sufficiently strong to make many Turks sceptical and suspicious of the motives behind the policies both of the EU and of individual European states. Perhaps a comparison can be made with Poland. That country, too, has a lot to gain, both economically and politically, from its integration in the EU, but historical factors, notably the memory of the division of Poland by big neighbours Germany and Russia, with the Poles themselves cast in the role of victims and tragic heroes, time and again lead the Poles to risk alienating fellow EU members through narrow nationalist policies, a course of action often as incomprehensible as the Turkish intransigence over Cyprus to outside observers with less acute historical memories. Turks are not Poles, and their position in Europe is very different, but the similarity perhaps rests in the importance of historical factors. Like human beings, states tend to act on the basis of what they think they remember about the past.

Greece. In Turkish eyes, the Turkish community in Cyprus falls into this category, and it is a national duty to protect it. Turks, as well as the public opinion in EU member states. Greece supported the policies of [illegible] in [illegible] and [illegible]. This [illegible] the [illegible] [illegible]

[illegible]

CHAPTER 5

Europe from the Balkans

Joep LEERSSEN

Athens, Istanbul, Odessa and Venice – no mean cities in European history. Between them they form an irregular quadrangle which, alongside the Iberian and Apennine peninsulas, is one of the three great southern European extensions into the Mediterranean. Named after the Balkans, one of the mountain ranges running across it, it is washed by the Adriatic, the Aegean and the Black Seas; less clearly defined is its land border to (and connection with) the European landmass to its north. The Sava and Danube rivers are sometimes used as a demarcation, but Romania (which lies north of the Danube) is generally considered a "Balkan country", while Slovene and Croatian areas south of the Sava strenuously resist and reject the appellation "Balkans" as somehow derogatory. (No such derogation is intended in what follows here; indeed my aim is, rather, to scrutinise and deconstruct such stereotyping imagery and discourse.) The true demarcation of the Balkans is not a matter of geography or "natural frontiers", but rather a matter of cultural and ideological connotations and vague historical echoes and reminiscences. The concept "Balkans" appears to exclude areas with a predominantly Catholic religious background (Slovenia, Croatia) and connotes the predominantly Orthodox regions of South-Eastern Europe with a history of lengthy Islamic, Ottoman domination. This gives an ambivalent in-between status to areas which historically were wholly or partly dominated by Venice, the Habsburgs or Russia: Dalmatia, the Voivodina, the Banat, Transylvania, Bucovina and Bessarabia. Across such border zones, which from Trieste to Odessa constitute perhaps the most intriguing belt of multi-ethnic and multi-cultural interaction areas in all of Europe, the "Balkans" proper form Europe's outrider to the Bosporus and Asia Minor.

In what follows here, we shall explore some of the implications of that geographical, geopolitical and historical fact. In doing so, we venture into what has recently become a flourishing scholarly field of interest. However, recent interest in the Balkans seems to have been triggered largely by the violent and tragic ethnic conflicts following the collapse of Communism in the region; some commentators go as far as seeing the ensuing hostilities as the re-emergence of an inborn tendency to ruthless, genocidal violence; while even the more moderate and balanced analysts use the recent clashes as a point of access to, and in some respect a lens on, the entire region's history.[1] The connotation of an allegedly characteristic tendency towards ruthless violence is, as imagologists have shown, one of Europe's ingrained prejudices and stereotypes concerning the region.[2] The extent to which the area has been seen through the filter of European stereotypes and prejudices has recently been explored by reputable scholars, among whom Vesna Goldsworthy, Ludmilla Kostova, Maria Todorova and Larry Wolff.[3] Ultimately, however, all such studies are concerned with the image of the Balkans in Western eyes; that is, with the Balkans as object, rather than as agent, of cultural attitudes. National self-images in Balkan countries have been explored by historians such as Andrew Wachtel and Rumen Daskalov,[4] but the extent to which Balkan self-images and

---

1 An intemperate book, obviously written under the impression of the atrocities of the Yugoslav Civil War, is Branimir Anzulovic, *Heavenly Serbia: from Myth to Genocide*, London, Hurst, 1999. Micha Glenny's benchmark book, *The Balkans: Nationalism, War and the Great Powers 1804-1999*, London, Granta, 1999, is a well informed, deservedly popular survey, but – like Robert D. Kaplan, *Balkan Ghosts: a Journey through History*, New York, St Martin's Press, 1993, or Marcus Tanner, *Croatia: a Nation Forged in War*, New Haven, Yale UP, 1997 – finalistic in seeing "ethnic hatred and war" as the overriding aspects, informing the principles and teleology of the region's history over the last two centuries. In fairness, Glenny's insightful and probing review essay, "Only in the Balkans", *London Review of Books* (29 May 1999), should be mentioned as a complement to his earlier book.

2 See the articles "Balkans" (by Ivana Živancevic-Sekeruš) and "Serbs" (by Pavle Sekeruš), in M. Beller & J. Leerssen (eds.), *Imagology*, Amsterdam, Rodopi, 2007, pp. 103-108 & 234-237, as well as the sources cited there.

3 Vesna Goldsworthy, *Inventing Ruritania: the Imperialism of the Imagination*, New Haven, Yale UP, 1998; Ludmilla Kostova, *Tales of the Periphery: the Balkans in Nineteenth-Century British Writing*, Veliko Turnovo, Veliko Turnove UP, 1997; Maria Todorova, *Imagining the Balkans*, Oxford, Oxford UP, 1997; Larry Wolff, *Inventing Eastern Europe: the Map of Civilization on the Mind of the Enlightenment*, Stanford, Stanford UP, 1994; and *idem*, *Venice and the Slavs: the Discovery of Dalmatia in the Age of Enlightenment*, Stanford, Stanford UP, 2001.

4 Rumen Daskalov, *The Making of a Nation in the Balkans: Historiography of the Bulgarian Revival*, Budapest, Central European UP, 2004; Andrew B. Wachtel, *Making a Nation, Breaking a Nation: Literature and Cultural Politics in Yugoslavia*, Stanford, Stanford UP, 1998. Also, Giorgos Agelopoulos, "Perceptions, Construction, and Definition of Greek National Identity in Late Nineteenth-Early Twentieth Century Macedonia", *Balkan Studies*, 36/2, 1995, pp. 247-263; Jelena Milojkovic-Djuric, *Panslavism and National Identity in Russia and in the Balkans, 1830-1880: Images of the Self and Others*, New York, Columbia UP, 1994; Janette Sampimon, *Becoming Bulgarian: the Articulation of Bulgarian Identity in the Nineteenth Century in its International Context, an Intellectual History*, Amsterdam, Pegasus, 2006.

identifications involved a positioning *vis-à-vis* the rest of Europe is still an under-researched area.[5] While, given the current author's linguistic limitations and the restricted scope of the present article, I cannot hope to address, let alone fill this lacuna, I wish at least to recommend it as a highly topical concern; in doing so, I draw on excellent regional histories by scholars like Barbara and Charles Jelavich and Traian Stoianovich,[6] but find it necessary to draw on literary and philological sources rather than the source material more usually studied by social and political historians. That point itself is, perhaps, indicative of the cleavage between political and cultural history concerning this European region.

Cultural attitudes surrounding, defining and positioning the Balkans in its relationship with the rest of Europe tend to conflate into an indiscriminate whole an area hugely diverse in religion and language. Who could trace the common denominators between the Albanian Bektaşi-Muslim Frashëri, the Montenegrin Prince-*Vladika* Petar Petrović Njegoš and the Croatian Archbishop Strossmayer, or between the Romanian folklorist-cum-statesman Vasile Alecsandri and the Greek insurgency leader Theodoros Kolokotronis? All of these are iconic, important figures in their respective countries' national nineteenth-century history, but the languages they spoke are mutually unintelligible, and their backgrounds and cultural contexts so wholly dissimilar that they represent, literally, different worlds. To elide all these differences and subsume them under a "Balkan" appellation ultimately derives from the region's one common, negative denominator: the Balkans is that part of the European continent where, from the fifteenth to the twentieth centuries, the Christian empires of the Habsburgs and (later) the Romanovs confronted the Islamic empire of the Ottomans. Battleground and interface, the Balkans is Europe's point of contact with the world of Islam –

5 Studies in travel writing are beginning to address the movements and perceptions in a more balanced mode; witness Wendy Bracewell & Alex Drace-Francis (eds.), *Balkan Travel Writing*, special issue 6/1-2, 2005, of *Journeys: the International Journal of Travel and Travel Writing*; and Nicolae Bocsan & Ioan Bolovan (eds.), *Călători Români in Occident, Secolele XVII-XX*, Cluj-Napoca, Institutul Cultural Român, 2004. An interesting study on mutual intra-Balkan ethnic perceptions is Nadia Danova, Vesela Dimova & M. Kalitsin (eds.), *Predstavata za 'Drugiya' na Balkanite*, Sofia, Marin Drinov, 1995.

6 Barbara Jelavich, *History of the Balkans: Eighteenth and Nineteenth Centuries*, Cambridge, Cambridge UP, 1983; Charles & Barbara Jelavich (eds.), *The Balkans in Transition*, Berkeley, University of California Press, 1963; Traian Stoianovich, *Between East and West: the Balkan and Mediterranean Worlds*, New Rochelle NY, Caratzas, 1992; *idem*, *Balkan Worlds: the First and Last Europe*, Armonk, NY, Sharpe, 1994; and Norbert Reiter (ed.), *Nationalbewegungen auf dem Balkan*, Wiesbaden, Harassowitz, 1983.

during the entire period, that is, when European modernity and the modern state developed.

The world of Islam has, for centuries, been Christian Europe's "Significant Other" – first in Spain (from Tariq ibn-Ziyad's invasion in 711 until the fall of Granada in 1492), and then in South-Eastern Europe. From the mid-fourteenth century onwards (well before the city of Constantinople finally fell to the armies of Mehmet II in 1453), the Ottoman Turks had first gained a foothold on the Thracian shore, then conquered large tracts of the Balkan peninsula, and with the fateful battle of Kosovo (1389) established their imperial encirclement of Byzantium. Serbia and the other Balkan lands became known as "Turkey-in-Europe". Even in the nineteenth century, travellers following the lure of Orientalism felt that they left the West, and entered the sphere of minarets and turbans, when leaving the Habsburg lands by crossing the Danube or the Sava into the Balkans. Although the Ottoman Empire at various periods stretched as far as Basra and Upper Egypt, and the Sultan also extended his influence over the Arabian peninsula and the holy places of Mecca, Medina and Jerusalem, the part of the Ottoman Empire that faced Europe was represented by its control of the Balkans, from where its shadow fell across Central Europe as far as Buda and the gates of Vienna.

To the extent, then, that Europe defined itself in Christian terms as that continent which was bordered by a hostile Islamic world, the Balkans was not really part of Europe. The fact that there were Christians living there under Ottoman rule also applied to Armenia and Georgia, and to the Copts, Maronites and Nestorians of the Middle East. Nor were even the Greeks themselves seen, throughout the seventeenth and for most of the eighteenth centuries, as "Europeans": they counted as a proverbially wily and mendacious Levantine trading nation living largely in a diaspora – much like Jews and Armenians. The famous Styrian "Tableau of Nationalities" of c. 1700, which displays European nations and their characteristics from Spain to Muscovy, has, at its extreme right-hand side, the figure of a turbanned Ottoman Levantine in baggy pantaloons, calling him indiscriminately "Turk or Greek" (*Tirk oder Griech*, Figure 5.1).[7] It was only with the classical philhellenism of Winckelmann, and with the romantic philhellenism of Byron, that Euro-

7 The original tableau is in the Austrian Ethnological Museum in Vienna; an in-depth iconographical and imagological analysis is given in F.K. Stanzel *et al.* (eds.), *Europäischer Völkerspiegel: imagologisch-ethnographische Studien zu den Völkertafeln des frühen 18. Jahrhunderts*, Heidelberg, Carl Winter, 1999; in that collection, specifically: Zoran Konstantinović, "'Tirk oder Griech'. Zur Kontamination ihrer Epitheta", pp. 299-314.

peans would come to see the Greeks as non-Levantines, as descendants of Aristotle and Pericles and as fellow-Christians.[8]

Figure 5.1. "Turk or Greek", from the Styrian "Tableau of Nationalities", c. 1700

Location: Österreichisches Museum für Volkskunde, Vienna.

This European view of the Balkans as "Turkey-in-Europe" applied the entire stock-in-trade of Orientalism to this region: sensuous

[8] On philhellenism: Loukia Droulia, *Philhellénisme. Ouvrages inspirés par la guerre de l'indépendance grecque, 1821-1833: Répertoire bibliographique*, Athens, Patris, 1974; Evangelos Konstantinou (ed.), *Europäischer Philhellenismus: die europäische philhellenische Literatur bis zur 1. Hälfte des 19. Jahrhunderts*, Frankfurt, Peter Lang, 1992; Alfred Noe (ed.), *Der Philhellenismus in der westeuropäischen Literatur 1780-1830*, Amsterdam, Rodopi, 1994; T. Spencer, *Fair Greece, Sad Relic: Literary Philhellenism from Shakespeare to Byron*, Athens, Harvey, 1986.

refinement and immoral hedonism linked to despotic rule and merciless cruelty.[9] The Christian population were seen as pauperised, oppressed helots, inhabitants of a rough mountainous terrain, obeying the primitive honour-and-shame code proper to mountain-dwellers: lawless, brutish and violent. What disappeared from view in this stereotype was the affluent mercantile and landowning Christian *millet* and their elite of Greek-Orthodox Phanariot rulers, who within the Ottoman system maintained a type of "Byzantium after Byzantium" (to recall the title of Nicolae Iorga's 1935 classic),[10] especially in the comparatively privileged and advanced Danubian principalities of Wallachia and Moldavia.

Figure 5.2. Dimitrie Cantemir (1673-1723)

Source: Dimitri Cantemir, *Extracts from "The History of the Ottoman Empire"*, ed. A. Duţu & P. Cernovodeanu, Bucharest, Association Internationale d'Études du Sud-Est Européen, 1973.

A telling case is that of the politician, diplomat and historian Dimitrie Cantemir (1673-1723), *Voivode* of Moldavia (and with princely titles also in the peerage of the Russian and Holy Roman Empires), author of the first European history of the Ottoman Empire (*Historia*

---

9 The literature on European Orientalism is vast. See the entry "Orientalism" by Bernd Thum in Beller & Leerssen (eds.), *Imagology*, pp. 389-393.

10 On Iorga: Maurice Pearton, "Nicolae Iorga as Historian and Politician", in D. Deletant & H. Hanak (eds.), *Historians as Nation-Builder: Central and South-East Europe*, London, Macmillan/SSEES, 1988, pp. 157-173.

Incrementorum atque Decrementorum Aulae Othomanicae, 1716), member of the Prussian Academy and co-founder of the St Petersburg Academy. A veritable intermediary between the European Enlightenment and Ottoman-dominated Europe, his achievements have somehow disappeared from view since they failed to fit into the schemata of the Balkan stereotype. Yet it was figures like him who shaped the peninsula's slow and difficult course towards modernity.

Figure 5.3. The Greek/Moldavian/Romanian Risings of 1821

Source: Private collection.

Nation building in the Balkan peninsula took place in the context of the long, slow decrepitude of the Ottoman Empire.[11] To some extent it was driven by the ambition of local magnates to occupy the power vacuum of the dwindling reach of the Porte (as in the case of Ali Pasha of Tebelen, the Njegoš Prince-Bishop of Montenegro, and, partially, Karadjordje in Serbia); it was also driven by an emerging middle class of merchants, professionals and intellectuals. Their organisation was, typically, non-territorial, taking shape in transregional networks and secret societies such as the famous *Filiki Eteria*. Such societies followed

---

11 I follow here the general historiography as surveyed above, and also the work of the project *Cultural nationalism in the Balkans during the nineteenth century: scholarly and intellectual institutions and networks in a multi-ethnic region* (cf. http://cf.hum.uva.nl/natlearn/balkan/index.html).

the template of the hermetic-masonic brotherhoods popular since the eighteenth century, often associated with subversive or progressive aims. As such, the *Filiki Eteria*, established by Phanariot Greeks in 1814, closely resembles the *Carbonari* in Italy. Numerous non-Greeks were members, and there is an important overlap between the Greek/ Moldavian/Romanian risings of 1821 (planned and carried by Ypsilantis and Tudor Vladimirescu, both members of the *Filiki Eteria*) (see Figure 5.3).

It is important to realise that such initiatives were largely non-territorial and "diasporic". They were centred in whatever metropolitan cities provided the points of congregation for like-minded spirits. Thus (to stay with the Greek example), the *Filiki Eteria* had its main centres in Odessa and Istanbul; West-European philhellenism was prepared by the activities of Korais in cities with Greek trading communities such as London, Paris and Amsterdam; the most important Greek-language periodical of the early nineteenth century, *O Logios Hermis*, was issued from Vienna and Bucharest.[12]

The fact that the early networks of national agitation and consciousness-raising were so very non-territorial – trans-national, trans-regional even – appears slightly anomalous to the modern reader, and accordingly the standard histories of the emerging nationalities of South-Eastern Europe have tended to downplay these aspects in favour of home-territory-based explanations. But it should be remembered that the Ottoman Empire was to a degree territorially indistinct. Its territorial-administrative division into *pashaliks* and *vilayets* had no correspondence whatsoever with the mingled ethnicities of the population, which were, instead, organised in the *millet* system of religious jurisdictions (for Muslims, Jews and Christians, regardless of region or language). Within such a structure, the distribution of national populations – or rather, the territorial distribution of a rising professional or mercantile middle class representing its nationality – need in no way reflect the outline of the states which became established on the former territory of Turkey-in-Europe.[13]

---

12 Polychronis Enepekides, "Beiträge zur geheimen geistigen und politischen Tätigkeit der Griechen in Wien vor der griechischen Revolution", *Berliner byzantinische Arbeiten*, 20, 1960; George Margaritis, "Griechenland: Wiedergeburt aus dem Geist der Antike", in Monica Flacke (ed.), *Mythen der Nationen: ein europäisches Panorama*, Munich, Koehler & Amelang, 1998, pp. 152-173; Vera Sýkora, "Vooruit, kinderen der Grieken…: Politisering en mythologisering van de Griekse nationale held Rigas Velestinlis Feraios", MA thesis, Universiteit Utrecht, History, 2005; and Marjolijne Janssen's contribution to the Athens workshop, Jan. 2005 of the Balkan Project, at http://cf.hum.uva.nl/natlearn/balkan/athens_janssen.pdf.

13 Raymond Detrez, "Het Osmaanse *millet*-systeem", in R. Detrez & J. Blommaert (eds.), *Nationalisme: kritische opstellen*, Berchem, EPO, 1994, pp. 290-303.

The Bulgarian case is particularly illustrative. Its earliest cultural consciousness-raisers were the monk Paisi based in the Mouth Athos monastery of Hilendar; a Pan-Slavist antiquary from the Ukraine, Yuri Venelin; the Heidelberg-educated physician Petar Beron, active most of his life in Paris; and the merchant Vasil Aprilov, who lived in Odessa. Paisi Hilendarksi wrote a history of the Bulgarians, which a half-century after his death in 1822 would become a national classic; Petar Beron (c. 1795-1871) wrote the first schoolbook primer in the Bulgarian language (significantly, while living in the towns of Braşov, present-day Romania); Yuri Venelin (1802-1839) developed an interest in Bulgarian culture while living in the Moldovan city of Kishinev (Chisinau), and propagated Bulgarianism in Moscow; and Vasil Aprilov worked and furthered Bulgarian education in Odessa. Significantly, all these activities took place outside what is nowadays Bulgaria; equally significantly, both Beron and Aprilov supported the Greek cause (possibly as sworn members of the *Filiki Eteria*) in their early years; there are even cases on record of nationally minded intellectuals who even now are unclassifiable as to their Bulgarian or Hellenic ethnicity, such as Nikolo Pikkolos. Similarly, the first Bulgarian-language newspaper was printed in Smyrna (present-day Izmir).[14]

While, then, the early consciousness-raising of the Balkan nationalities was territorially free-floating, urban and metropolitan centres could become centres of many different national movements. Notoriously, the *Stammtisch* or "regulars' table" of the Vienna-based Slovene intellectual Jernej Kopitar, at the inn *Zum weissen Wolfen*, became a meeting place for minority nationalities of all the Habsburgs' Balkan dominions, as well as for exiles from the Ottoman-dominated lands. The national-romantic German poet Hoffmann von Fallersleben (author of the *Deutschlandlied*) even dedicated a poem to this international nexus of nationalist conviviality.[15] Vienna, Odessa and, above all, Bucharest with its regionally important lyceum, were similar meeting-places for emerging national activists: the first alphabet for the Albanian language was devised there by Romania-based Naum Veqilharxhi.[16]

---

14 Sampimon, *Becoming Bulgarian*; Nikola Kolev, "La collaboration entre le Dr Nicolas S. Piccolos et Claude Fauriel", *Etudes Balkaniques*, 2 (1978), pp. 114-123; Albert B. Lord, "Nationalism and the Muses in the Balkan Slavic Literature in the Modern Period", in C. & B. Jelavich (eds.), *The Balkans in Transition*, pp. 258-296.

15 Ingrid Merchiers, "Cultural Nationalism in the South Slav Habsburg Lands in the Early Nineteenth Century: the Scholarly Network of Jernej Kopitar (1780-1844)", doctoral thesis, Universiteit Gent, Faculteit Letteren en Wijsbegeerte, 2005.

16 Myslim Islami, *Naum P. Veqilharxhi*, Prishtinë, Rilindja, 1969; Robert Elsie, *History of Albanian Literature*, 2 Vols, New York, Columbia UP, 1996.

The beginnings of anti-Turkish nationalism in the Balkans were, then, territorially indistinct networks of intellectuals ramifying across the peninsula, and beyond into Southern Russia, Anatolia, Venice and Central Europe. The intellectuals concerned, for their part, certainly did not consider Europe as a "foreign country", but rather as the natural hinterland of their geopolitical and cultural ambitions. The Serbs exploited a position straddling the Danube, with one leg in the Ottoman *pashalik* of Belgrade and the other in the Habsburg-dominated region of Voivodina. Armed resistance occurred in the Belgrade *pashalik*, while cultural consciousness-raising used the printing presses of Venice, Buda and Vienna. The first Serb reading society or *matica* was established among the Serb community of Buda, and later relocated to the capital of Voivodina, Novi Sad. It was here that contacts were fostered with Croats, in a climate of pan-Slavic solidarity presided over by the great Slovak antiquary and philologist Pavol Šafárik, who taught at the Novi Sad lyceum. These contacts enmeshed Serb nationalism in the Austro-Slavic agenda of the Croatia-based "Illyrian movement", and were to culminate in 1850 in a joint statement by Croat and Serbian authors gathered in Vienna and proclaiming the unity of the "Serbo-Croat" literary language.[17]

Meanwhile, the great Serbian philologist Vuk Karadźić, who had taken refuge from insurrectionary and repressive violence in Vienna, had put the Serb language on the literary map of Europe by publishing collections of Serb folksongs (hailed by contemporary philologists like Jacob Grimm and Claude Fauriel as a form of "oral epic"), by publishing a Serb dictionary and grammar, and by reforming the Serb alphabet, thus codifying and individuating as a language something which until then had merely been seen as a debased oral offshoot of the liturgical language of Church Slavonic. Karadźić's work, though based on the author's broad and deep knowledge of his native culture and carried by his patriotic zeal and his erudition, was facilitated and directed under the guidance and patronage of the aforementioned Jernej Kopitar, a senior official at Vienna's Imperial Library, and one of the foremost Slav

[17] Źivan Milisavac, *Matica Srpska, 1829-1964*, Novi Sad, Matica Srpska, 1965; Elinor Murray Despatalovic, *Ljudevit Gaj and the Illyrian Movement*, Boulder, East European Quarterly, 1975; Branko Franolić, "The Development of Literary Croatian and Serbian", in I. Fodor & C. Hagège (eds.), *Language Reform: History and Future*, Hamburg, Buske, 1983-94, pp. 85-112; Josef Hamm, "Randbemerkungen zur Entstehung und Entwicklung der serbokroatischen Schriftsprache", in C. Hannick (ed.), *Sprachen und Nationen im Balkanraum: die historischen Bedingungen der Entstehung der heutigen Nationalsprachen*, Vienna, Böhlau, 1987, pp. 65-76.

scholars of his generation with excellent contacts in German intellectual circles.[18]

As all these names and cases show, Europe was, for the emerging intelligentsia on the Balkans, a sounding board; by no means an Other. In the northern parts of the Balkans, we see a general participation in Europe's Enlightenment climate of disinterested scholarship in the service of the Fatherland's reputation and the advancement of learning. The early cultivation of the Romanian language and tradition in Transylvania, and of Croat and Slovene, in regions under Habsburg rule (directly or by way of the ancillary capital of Buda/Pest), matches the comments made concerning the position of Serb in the Voivodina. There was, then, an osmosis whenever South-East European nationalities straddled the frontier between the Habsburg and Ottoman Empires; even the geographically isolated lands where Greek and Bulgarian were the vernacular language had contacts with a European hinterland by way of diaspora networks. The case of Montenegro also fits this pattern, although that country shared no territory or population with "European" neighbours. Its Prince-Bishop Petar Petrović II Njegoš, now often remembered for his harshly, even murderously anti-Turkish poetry, also wrote Enlightenment-style verse on encyclopedic themes, behaved like an Enlightened autocrat in his small realm, and followed in the footsteps of educational and encyclopedic educationalists who had been active in the region in the course of eighteenth century.[19] Even in the case of Albanian culture, there was a European connection: that of the Arberësh community long settled in southern Italy, which caught the climate of the romantic-national *Risorgimento*, and celebrated its Albanian roots and traditions in literary texts.

---

18 Merchiers, *Cultural Nationalism*. The literature on Karadžić is very substantial; see especially Vera Bojic, *Jacob Grimm und Vuk Karadžić: ein Vergleich ihrer Sprachauffassungen und ihre Zusammenarbeit auf dem Gebiet der serbischen Grammatik*, Munich, Otto Sagner, 1977; and Duncan Wilson, *The Life and Times of Vuk Stefanović Karadžić, 1787-1864: Literacy, Literature and National Independence in Serbia*, Oxford, Clarendon Press, 1970.

19 Examples in point are Baron Žiga Zois (Trieste 1747 – Ljubljana 1819), naturalist, patron of learning and employer of young Kopitar (cf. Merchiers, *Cultural Nationalism*, and Petra Testen, "Žiga Zois: Biografska skica", *Glasnik Slovenske Matice*, 29-31, 2005-07, pp. 118-125). Also, the Croatian Franciscan monk Andrija Kaćić-Miošić, 1704-60, author of a popularising-educational national history, the *Razgovor ugodni naroda slovinskoga*, 1756; or the Slovene/Austrian Johann Sigmund Popowitsch, 1705-74, grammarian, dialectologist and author of an encyclopedic/philological treatise *Untersuchungen vom Meere*, 1750.

Figure 5.4. Dora d'Istria (1828-1888)

Source: Marquis de Villemer, *Les portraits cosmopolites*, Paris, 1870.

All these cross-currents are exemplified in the figure of the colourful Dora d'Istria, born Alexandra Ghica (1828-1888). A writer and cosmopolitan socialite of Europe-wide fame (and the first woman, incidentally, to climb Mont Blanc), she was a scion of the noble Ghica family, Phanariot aristocracy with a long record of public service in the Danubian principalities. Dora, herself a native speaker of Albanian, became a protector of many an Albanian or Arberësh man of letters and, in her way, the Mme de Staël of the Balkans.[20]

Balkan nation building did get caught up, however, in the European geopolitics around Russophile Panslavism. The decay of the Ottoman "Sick Man of Europe" left a power vacuum between Italy and the Middle East in which all European powers saw their interests implicated – foremost among them the Habsburg and Romanov empires. In the long run-up to the shooting incident at Sarajevo in 1914, a fierce struggle for political and territorial leverage had placed the region under a heavy mortgage. The fact that various Balkan countries achieved independence in the course of the nineteenth century was itself something which Metternich's Congress system had not allowed for. A precarious

20 Elsie, *History of Albanian Literature*; Antonio D'Alessandri, "'La vie monastique dans l'Eglise orientale' di Dora D'Istria (1828-1888): problematiche religiose e politiche", *Trimestre*, 36/1-2, 2003, pp. 73-92.

balancing act was maintained in that the sovereigns of the new states (which, owing to the prevailing Orthodox religion, listed towards Russian influence) were of German stock – Wittelsbachs, Saxe-Coburg-Gothas and Hohenzollerns.

Within Europe at large, the nationalities problem was brought to a fine point in the dilemmatic split, in the Pan-Slavic movement, between an Austro-Slavic trend (looking to Vienna as the centre of European Slavdom) and a Russophile wing.[21] Vienna had the greatest number of Slavic nationalities under its aegis, Moscow the greatest number of Slavs. Those who adhered to a "civic" form of nationalism were mistrustful of the Tsar's autocracy; that trend predominated in those Slavic nations traditionally oriented towards Latin Christianity (as opposed to Orthodoxy), so that, barring a few exceptions,[22] the split between the Austro-Slavic and Russophile wings of Pan-Slavism corresponds roughly with the division between Western and Eastern Christianity.

In the Balkans, that same split has (again, roughly) a North-South dimension. Serbs, Montenegrins, Macedonians, and Bulgarians tended on the whole to look to the Tsar as their main European ally, a gravitational pull which was far less pronounced among their neighbours to the North – Croats and Slovenes or even the Voivodina Serbs. However, the neatness of that division is shot through by the attitudes among the non-Slavic Orthodox nations of the Balkans (Greece, Romania) whose religious Orthodoxy aligned them with a Russophile tendency, but who also evinced a more pronounced

21 Generally, Hans Kohn, *Pan-Slavism, its History and Ideology*, New York, Vintage, 1960.

22 The cases of Slovakia and the Ukraine are anomalous. The 1848 generation of the Ukraine, around Taras Shevchenko, were Eastern/Orthodox in their situation, but liberal and "civic" in their orientation; conversely, the group of Slovak nationalists around Ljudovit Štúr were so apprehensive of Czech cultural hegemonism that they were more Russophile. Cf. Ralph Lindheim & Georges S.N. Luckyj (eds.), *Towards an Intellectual History of the Ukraine: an Anthology of Ukrainian Thought from 1710 to 1995*, Toronto, Toronto UP, 1996; Pavlo Zaitsev, *Taras Shevchenko: a Life*, trl. G.S.N. Luckyj, Toronto, Toronto UP, 1988; Peter Brock, *The Slovak National Awakening: an Essay in the Intellectual History of East Central Europe*, Toronto, Toronto UP, 1976; Ján Buncák & Magdalena Piscová, "Modern National Identity of Slovaks and their Attitude towards Europe", *Sociologia*, 32/3, 2000, pp. 289-310; Ismo Nurmi, *Slovakia: a Playground for Nationalism and National Identity: Manifestations of the National Identity of Slovaks, 1918-1920*, Helsinki, Suomen Historiallinen Seura, 1999; David Short, "The Use and Abuse of the Language Argument in mid-Nineteenth Century 'Czechoslovakism': an Appraisal of a Propaganda Milestone", in R.B. Pynsent (ed.), *The Literature of Nationalism: Essays on East European Identity*, London, Macmillan/SSEES, 1996, pp. 40-65.

Westward leaning in their metropolitan cultural and political matters.[23] The Romanian shift from the Cyrillic to the Latin alphabet in the mid-nineteenth century emphasised the increasingly important realisation that the Romanian language belonged to the Romance family, despite its East-European location; the poet/philologist and statesman Vasile Alecsandri cultivated his country's links with France, culminating in the *fin de siècle* reputation of Bucharest as the "Paris of the Balkans". With artists like the poet Eminescu, the composer Enesco and the sculptor Brancusi, Romania developed into a fully fledged participant in the high culture of European metropolitan modernity. The same, indeed, goes for Greece – witness the figure of a composer like Kalomiris.[24] Although, in the manner of most "new" nation states, Greece cultivated its Hellenic roots and sense of identity, Athens is in every sense of the word a post-imperial European metropolis much like Helsinki, Budapest, Oslo or Dublin. Athens and Bucharest share with the rest of Europe a characteristic urban architecture and a city-scape of nineteenth-century vintage with a university, a national theatre, a national museum, a *conservatoire*, a public transport system, and all the other trappings. In the light of this realisation, perhaps one needs also to tone down the opposition between a Russian and a Western tendency. Russian high culture at this time was, after all, that of Fabergé eggs, Tchaikovsky's *Swan Lake*, Turgenev and the trans-Siberian railroad, rather than Rasputin; and the Orient Express was no longer just a destination into the exotic, but rather a modern piece of engineering bringing far-flung parts of Europe closer together.

In this sense, too, the Balkans should not be seen as Europe's Other or outside – not even the greatest Balkan city of all, Istanbul, which was to a significant extent participating in this modernisation process. The nation-building process in the Balkans was made possible by, and took place in the context of, a modernisation process which affected all of Europe, from Reykjavík to Helsinki and from Barcelona to Odessa. The national sense of identity which was cultivated as part of this process

---

23 Similarly, in the Albanian national movement, traditionally a strong Catholic and Italy-oriented presence in the North coexisted with Islamism and Bektaşi attitudes in the South as evinced by the extraordinary Frashëri brothers. Cf. generally Elsie, *History of Albanian Literature*.

24 Viorica Niscov, "Über den Widerhall der volkskundlichen Beschäftigung der Brüder Grimm in Rumänien", *Brüder Grimm Gedenken*, 2, 1975, pp. 146-167; Klaus-Henning Schroeder, "Marginale Romania: die Bildung und Normierung der rumänischen Schriftsprache", in C. Hannick (ed.), *Sprachen und Nationen im Balkanraum: die historischen Bedingungen der Entstehung der heutigen Nationalsprachen*, Vienna, Böhlau, 1987, pp. 195-209; Olympia Frangou-Psychopedis, *I ethniki scholi mousikis: Provlimata ideoloyias*, Athens, Idryma Mesogeiakon Meleton, 1990.

was cultivated by a transnational network of intellectuals who firmly had their eye on Europe as their natural hinterland. If anything, the remarkable thing about Balkan nationalism is that it turned its face to the North, and that means: away from the myth of a Constantinople Redeemed.

The First World War did not, in a sense, start in 1914; it was really a scale enlargement of the Balkan Wars which had been raging in 1912-13 between the states newly established on the ruins of the crumbling Ottoman Empire. The Balkan Wars were fought with a heavy geopolitical investment from the European powers, and provided the tripwire of the Sarajevo assassination. From the very beginning of the twentieth century, therefore, the Balkan nation states, which had asserted their identity and self-determination on the basis of a shared European identity, came to see the European powers increasingly as hegemonic meddlers in their affairs.

The European powers, for their part, continued a habit of exoticising Europe's south-eastern peninsula. For them, it was theatre of war carved up into operetta states and inhabited by gypsies, brigands, semi-Oriental mountaineers and flamboyant warlords. Anthony Hope's adventure romance *The Prisoner of Zenda* (1894) and Franz Léhar's operetta *Die lustige Witwe* (1905) – and even G.B. Shaw's *Arms and the Man* (1894) – continued stereotypes at least as old as Byron's encounter with Ali Pasha at Ianina.[25] Fictional names have proliferated for this marginal never-never-land of the European imagination. The name "Ruritania" has swirled around from *The Prisoner of Zenda* to Ernest Gellner's *Nations and Nationalism* (1983), by way of P.G. Wodehouse.[26] Other coinages include "Pontevedro" and "Marshovia" (in Léhar's *Die lustige Witwe* and its 1934 film version, *The Merry Widow*); the neighbouring states of "Syldavia" and "Borduria" (in Hergé's *Tintin* comics); "Slaka" (in Malcolm Bradbury's novel *Rates of Exchange* and its spin-off, the spoof travel guide *Why Come to Slaka?*); and "Molvania" in the pas-

25 The encounter occurred in 1809 and was made famous in Byron's *Childe Harold*. Ali Pasha also featured in one of the many sub-plots of Alexandre Dumas's (heavily Byronic) *Le Comte de Monte-Cristo* (1844). Cf. my "Een omzwerving rond Monte-Cristo. Mediterrane beeldvorming van Byron tot Bogart en Braudel", in A.M. van Erp Taalman Kip & I.F. de Jong (eds.), *Schurken en schelmen: cultuurhistorische verkenningen rond de Middellandse Zee*, Amsterdam, Amsterdam UP, 1995, pp. 9-24.

26 Goldsworthy, *Inventing Ruritania*. The name was also used for a fictional Balkan country in Evelyn Waugh's *Vile Bodies*, 1930.

tiche travelogue *Molvanîa: a Land Untouched by Modern Dentistry* (2005).[27]

Conversely, the very real region of Transylvania and the very real mountain range of the Carpathians have in the mind of Western readers become fictional or supernatural, make-believe place-names, after having become the setting of Bram Stoker's *Dracula* and its numerous derivatives. Thus the Balkans has, from the geographical margin of Europe, been pushed to the ontological margins of the European imagination, located somewhere between the real world and implausible fantasy.

All this bespeaks a vertiginous cognitive mismatch, which continued well into the 1990s: the Balkans sees itself as part of Europe, while Europe persistently exoticises it. Tragically, Europe has turned itself into the Balkans' Other by its combination of uncomprehending exoticism and high-handed interference. There is a word for this: alienation.

---

[27] 1905. For a full list, see http://en.wikipedia.org/w/index.php?title=Fictional_European_countries&oldid=173916022.

PART IV

FROM OUTSIDE EUROPE

CHAPTER 6

"I Would Rather Go to Europe than Go to Heaven"

Images of Europe in the United States[1]

Ruud JANSSENS

Americans have been fascinated by European culture throughout their history. Before the American Impressionist painter William Merritt Chase ever went to Europe, he already had exclaimed, "My God, I would rather go to Europe than go to Heaven".[2] In contrast to Chase's overly optimistic image of Europe there was an opposing notion about the old continent, as expressed by Mark Twain in *Innocents Abroad*, his ironic travel account of a group of Americans visiting Europe and the Middle East. In Paris, Twain wrote, "We went to see the Cathedral of Notre Dame. – We had heard of it before. It surprises me, sometimes, to think how much we *do* know, and how intelligent we are. We recognised the brown old Gothic pile in a moment; it was like the pictures".[3]

Chase and Twain exemplify the complex nature of the relationship between America and Europe. Twain wanted to show how naïve Americans could be about Europe and yet how superior they could feel, while also having a need to go to Europe to admire higher culture – in fact expressing a notion of inferiority. In a way, Twain and Chase were stating the same ideas about Europe: Chase's optimism was only possible because of the one-dimensional view of Europe which Twain emphasised. He also emphasised that American images of Europe were a reflection not only on Europe but also on American society. Chase's optimism about Europe simultaneously expressed pessimism about American culture.

1 I would like to thank Barbara Weber, Charlotte Hille, and Kate and Tom Delaney for their suggestions and comments on an earlier draft of this essay. I am responsible for any remaining errors.

2 Chase quoted in Kathleen Adler, "'We'll Always Have Paris': Paris as Training Ground and Proving Ground", in Kathleen Adler, Erica E. Hirshler, H. Barbara Weinberg, *Americans in Paris, 1860-1900*, London, National Gallery, 2006, p. 14.

3 Mark Twain, *Innocents Abroad & Roughing It*, New York, Library of America, 1984, p. 104.

Since the early seventeenth century, Europeans have migrated to North America, regularly and in large numbers. European nations have had colonies on the North American continent, in Middle and South America – and still have them in the Caribbean. The relations between Americans and Europeans – economic, political, social, cultural, and personal – are countless. A focus on some key images that Americans have held about Europe will elucidate the richness, complexity, and sometimes ambivalence of the relationship.

In 1947, Professor of American Studies A.N.J. den Hollander stated in his inaugural lecture, "As Others See Us", that when people form an opinion about another nation, they choose from a "multiplicity of notions already existing about what is living beyond the frontiers".[4] In her study of American images of the Japanese, Sheila Johnson used the same idea. She found that Americans have negative and positive stereotypes of Asians, and that choices among stereotypes depended on the political situation. This essay presents a number of basic images Americans have of Europeans.

As Johnson found with Asia, American images of Europe can easily change from negative to positive. A good example is the recent debate about the war with Iraq. In a Pew opinion poll of February 2002, seventy-nine per cent of Americans were positive about France, while sixteen per cent were negative. In a June 2003 poll, after France had condemned the American-led invasion of Iraq, sixty percent of Americans judged France unfavourably and only twenty-nine percent favourably.[5] As the British have found, a country's "popularity" among Americans can easily shift with time, but such fluctuations do not preclude a positive and close relation with the United States. In spite of its origin as a former British colony, the War of 1812, and the unpopularity of Britain in the United States during the Boer War, the United States still developed their "special relationship" with Great Britain.[6]

One basic image Americans continued to have from the outset is that the United States would be different from Europe. The United States would not have the political or religious oppression that plagued the old continent. This ideal was as true for the Puritans, escaping Europe and trying to create their "city upon a hill," as it was true for the Founding

4 A.N.J. den Hollander, "As Others See Us", *Synthese*, 6/5&6, May 1947-May 1948, pp. 214-237.

5 The Pew Global Attitudes Project, *Views of a Changing World*, at http://people-press.org/reports/display.php3?ReportID=185, consulted 25 Sept. 2007.

6 See Christopher Hitchens, *Blood, Class and Empire: the Enduring Anglo-American Relationship*, New York, Nation Books, 2004.

Fathers defining a new political system. Historian Richard Pells pointed out what he believed was the key message in this reaction to Europe:

> America would be different, exceptional, a place of infinite opportunity and possibility for all who settled there. A vast untapped and unspoiled continent beckoned. Americans were embarked on a special errand into the wilderness. For eighteenth-century colonists and nineteenth-century immigrants, such notions led inevitably to an assertion of superiority. [...] America embodied innocence, youthfulness, vigor, confidence, optimism, freedom, and (once the wilderness was conquered) prosperity and modernity. Europe represented deviousness, cynicism, corruption, decadence, fatigue, poverty, social and ideological conflict, war. This type of discourse – self-congratulatory, heavily moralistic, serene in its conviction that America was good and Europe was evil – had grown familiar by the early nineteenth century and remained central to America's image of itself as a real and symbolic alternative to Europe through much of the twentieth century.[7]

Yet, as Pells also pointed out, Europe not only appeared decadent to Americans but also was a source of inspiration in regard to "high" culture, a point to which I will return.

In order to offer an insight into the stereotypes of American views toward Europe, ranging from rejection to inspiration, I will focus on three main themes. First, the political theme concerns the strategic aspects of the relations between the United States and Europe, in which the United States tried to deal with the bellicose nature of European politics. Second, the social theme concerns migration. The majority of American citizens are of European origin. These European roots could lead to a certain nostalgia about the home country and toward people of the same national origin. Yet, if the reason for departure from Europe had political, religious, or economic causes, there was less reason to think in positive terms about the Old World. Third and last, culture played an important role in the perception of Europe by Americans. In the nineteenth century and the beginning of the twentieth, Americans admired European culture. In the second half of the twentieth century, American culture played a dominant role in Europe (and elsewhere), while Europe, the Old World, appeared increasingly "quaint", to put it positively, for Americans. Finally, in addition to high-brow culture such as literature and painting, I will also look at popular culture, as expressed in a sample of Hollywood movies in which Americans visit Europe, making it possible for the characters and movie makers to comment directly on European culture and society.

---

[7] Richard Pells, *Not Like Us: How Europeans have Loved, Hated, and Transformed American Culture since World War II*, New York, Basic Books, 1997, p. 3.

A major challenge in writing about American images of Europe is to define the level of generalisation on which images occur. Even if Americans regularly spoke of Europe, they often meant a specific nation or even city. If there is no "Europe," it is also complicated to write about "the United States" in this context. It is important to note which Americans gave their views on Europe and when. Wealthy, well established, New England, nineteenth-century families, for instance, had different perceptions of Europe from second or third-generation immigrants after World War II.

A Continent of Warring Nations: American Images of Political Europe

From the very founding of their nation, United States politicians have seen Europe as an inspiration and as a concern. The American political system was based on European examples, ranging from ancient Athens and Rome to the ideas of Renaissance Italy, the French Enlightenment, and British political traditions.[8] The Founding Fathers did not see European politics only as a positive example. American politicians were also concerned about the many wars in Europe and possible American involvement in these conflicts. After participating in World War I (1917-1918) and World War II (1941-1945), American political leaders committed themselves to European co-operation and integration, in the hope that their support for a more unified Europe would decrease the chances of yet another major war.

In the Declaration of Independence, the American political leaders emphasised the wrongdoings of King George III. Their new political system would be a republic. The Founding Fathers looked at various European republics as examples and inspirations for their own government. The Polish and Dutch Republics were seen as inefficient systems: ancient Greece and Rome were much more admired. In the case of Rome, the Roman Republic was especially inspiring, while its decline into the Roman Empire was taken as a warning. The Roman Republic as a political inspiration returned in many guises. In the newly built capital Washington, DC, the architecture of many government buildings was classical, and the local Goose Creek was renamed the Tiber.

The newly created political system in many ways reflected closeness to Europe. Yet, concerning politics, and in particular foreign policy,

8 See Carl Richard, *The Founders and the Classics: Greece, Rome, and the American Enlightenment*, Cambridge MA, Harvard UP, 1995; J.G.A. Pocock, *The Machiavellian Moment: Florentine Political Thought and the Atlantic Republican Tradition*, Princeton, Princeton UP, 1975; and Gordon S. Wood, *The Radicalism of the American Revolution*, New York, Vintage, 1993.

American politicians wanted to keep a certain aloofness in regard to the Europeans. President George Washington had been a major in the British Army during the French and Indian War (1754-1763; in Europe also known as the Seven Years War (1756-1763)). Washington's unit was actually involved in starting the conflict, which resulted in a war around the world. This conflict showed how the fate of American colonists was related to the geopolitical interests of, in this case, the British and the French. As commander of the American forces during the War of Independence and his Presidency, Washington became convinced that the United States should keep its distance from the European powers, if it wanted to stay out of major wars. As he said in his Farewell Address,

> The great rule of conduct for us in regard to foreign nations is in extending our commercial relations, to have with them as little political connection as possible. So far as we have already formed engagements, let them be fulfilled with perfect good faith. Here let us stop. Europe has a set of primary interests which to us have none; or a very remote relation. Hence she must be engaged in frequent controversies, the causes of which are essentially foreign to our concerns. […] Why, by interweaving our destiny with that of any part of Europe, entangle our peace and prosperity in the toils of European ambition, rivalship, interest, humor or caprice? It is our true policy to steer clear of permanent alliances with any portion of the foreign world […].[9]

At the end of the eighteenth century, Washington's concern was justified in part by the presence of British, French, and Spanish colonies on the North American continent.

During the nineteenth century, all American presidents took Washington's advice as their guideline (even though the French and Spanish colonies disappeared, as did the later Russian colony in Alaska). European nations, as the most powerful countries globally, continued to determine the foreign policy of the United States. In 1823, Secretary of State John Quincy Adams drew up the Monroe Doctrine (named after President James Monroe) to warn European powers to keep out of the Americas. Even though the policy only worked because of the shared interest of Great Britain and the United States in the region (with the British having the leading role), the doctrine stresses American concern about possible imperialistic policies of European powers.

9 President George Washington, *Farewell Address*, 1796. An excellent study of the origins and early years of United States foreign policy remains Felix Gilbert, *To The Farewell Address: Ideas of Early American Foreign Policy*, Princeton, Princeton UP, 1961.

The same concern drove the so-called Open Door Policy in China. At the end of the nineteenth century, it seemed that European powers were to divide the Chinese Empire among themselves. In 1899, Secretary of State John Hay proposed a policy of equal economic rights in China for all (foreign) powers involved. The principle of equal access to China was important to the United States, which had trade interests in China as well (the United States was also in the middle of establishing control over the Philippines and had already obtained colonies in the Pacific Ocean as coal stations on the way to China). The European powers supported the Open Door Policy – and again Britain played an important role – because they did not want to go to war with each other or, potentially, with the United States over China.

The two World Wars in the first half of the twentieth century changed the strategic orientation of the United States. Where George Washington had argued for non-entanglement with Europe (permitting trade relations, but no military alliances), American political leaders came to realise that even if the American public wanted to stay out of these "European" wars, eventually the United States would have to get involved in any case. President Woodrow Wilson tried to create an international political structure with his Fourteen Points, which sought to make a future conflict in Europe less likely.[10] However, he did not get the support of the United States Congress to participate in the League of Nations.[11] At the end of World War II, President Franklin D. Roosevelt and his successor Harry S. Truman were able to establish the United Nations. Most Americans realised that the weapons now used in war, such as submarines, aircraft carriers, long-range bombers, missiles, and atomic bombs, made it less likely that the Atlantic and Pacific Oceans would protect the United States from direct enemy attacks on the American homeland (in that context the attack on Pearl Harbor by the Japanese forces in 1941 played an important role).

Truman realised that relief aid and economic support to Europe and Asia were crucial to American interests, both for the American economy and its national security. As Truman wrote in his memoirs: "To help peoples in distress was not only a tradition of our country but was also

---

10 See Thomas J. Knock, *To End All Wars: Woodrow Wilson and the Quest for a New World Order*, Princeton, Princeton UP, 1992.

11 There were other problems with Wilson's ideas as well, such as that the self-determination of people was only intended for the people within the Habsburg empire (and not for the Vietnamese in the French empire, for instance, as Ho Chi Minh tried to argue (see Margaret MacMillan, *Paris, 1919: Six Months that Changed the World*, New York, Random House, 2001, p. 59)), and that the League of Nations did not have an armed force to enforce its decisions (as the Japanese revealed with their conquest of Manchuria in 1931, against the wishes of the League).

essential to our security. By rebuilding Europe and Asia, we would help to establish that healthy economic balance which is essential to the peace of the world".[12] With the onset of the Cold War, Truman decided that the long standing policy of non-entanglement with Europe should be changed, and a military commitment was necessary to support the essential alliance with Western Europe in this new conflict:

> The Marshall Plan had brought some relief, but the constant threat of unpredictable Soviet moves resulted in an atmosphere of insecurity and fear among the peoples in western Europe. Something more needed to be done to counteract the fear of the peoples of Europe that their countries would be overrun by the Soviet Army before effective help could arrive. Only an inclusive security system could dispel these fears.[13]

This security system became the North Atlantic Treaty Organisation (NATO).

Even though United States policy toward Europe changed from non-entanglement to a military alliance over a period of two centuries, it could be said that the basis of American policy remained the same: Europeans fight wars that eventually involve the United States. European diplomacy was regularly frustrating for American policy makers. The remark made by Secretary of Defense Donald Rumsfeld about the "old" and the "new" Europe, when France and Germany did not want to support the war in Iraq in 2003, is just a recent example. In 1973, President Richard Nixon and his National Security Advisor Henry Kissinger tried to strengthen ties between the United States and Europe, but with limited success. As Kissinger stated, "The United States has global interest and responsibilities. Our European allies have regional interests. These are not necessarily in conflict, but in the new era neither are they automatically identical".[14] Kissinger was also frustrated by the lack of consensus and unity in European politics, as he implied with his question, "What's Europe's phone number?".[15] Then again, it could be argued that it was often easier for Americans to talk about Europe – precisely because they had more emotional distance, or less intimate knowledge of the Continent – than it was for many French, Italian, German, or British citizens to think in terms of what they had in common.

---

12 Harry S. Truman, *Memoirs: Years of Trial and Hope, 1946-1952*, New York, Signet Books, 1956, p. 135.

13 Truman, *Memoirs*, p. 286.

14 Henry Kissinger, *Years of Upheaval*, Boston, Little Brown, 1982, p. 153.

15 Although the quote is frequently used, the phrasing is always almost different, and it is hard to track down the original wording. See "Dr. Henry Kissinger at the second SHAPE lecture series", at http://www.nato.int/shape/multi/video/2005/lecture/kissinger/s051104a.htm, accessed 10 Nov. 2005.

This distrust of European politics and warfare was present during the presidency of Bill Clinton as well. After initially keeping out of the conflicts concerning the former Yugoslavia, Clinton decided that the United States should get involved in the Bosnia crisis in 1995. As he wrote to Speaker of the House Newt Gingrich,

> This Administration, and that of previous Democratic and Republican Presidents, have been firmly committed to the principle that the security and stability of Europe is of fundamental interest to the United States. The conflict in Bosnia is the most dangerous threat to European security since the end of World War II. If the negotiations fail and the war resumes, as it in all probability would, there is the very real risk that it could spread beyond Bosnia, and involve Europe's new democracies as well as our NATO allies. Twice this century, we paid a heavy price for turning our backs to conflict in Europe.[16]

Secretary of State Warren Christopher had made a similar argument earlier, contending, "Twice in this century we have had to send our soldiers to fight in wars that began in Central Europe".[17]

This conviction about the consequences of a possible war in Europe also led the Clinton administration to support an enlarged European Union, including East European nations. The Clinton administration hoped that a further integration of Europe, economically and politically, would prevent the threat of conflicts in the long run. When the member states of the European Union were not willing to include members of the former Soviet bloc in their organisation, the Clinton administration decided to expand NATO and by that method pave the way for an eventual expansion of the European Union. Clinton's strategy proved correct, and successful co-operation in NATO was one argument which the West European partners could use in order eventually to include those same nations in an expanded European Union.[18]

A Nation of Immigrants

While nineteenth-century American presidents tried to stay out of European power politics, the American population was confronted with European immigrants in large numbers. The descendants of the original

---

16 President Clinton quoted in Ted Galen Carpenter, "The Domino Theory Reborn: Clinton's Bosnia Intervention and the 'Wider War' Thesis", *Cato Foreign Policy Briefing,* 42, 15 Aug. 1996, at http://www.cato.org/pubs/fpbriefs/fpb-042.html, accessed 25 Sept. 2007.

17 Warren Christopher, quoted *ibid.*

18 See Ron Asmus, *Opening NATO's Door: how the Alliance Remade itself for a New Era*, New York, Columbia UP, 2002.

European immigrants considered themselves to be "native Americans".[19] Immigration from traditional countries of origin, such as Great Britain or Germany, continued. Meanwhile, from other parts of Europe, large groups of new immigrants moved to the United States as well. In a number of ways, they were considered to be different. The Irish, entering the United States since the 1830s, were Catholic, poor, and viewed as unhygienic. According to the first census in the United States, in 1820, there were 9,638,000 Americans, of which twenty per cent were African American. Between 1847 and 1854 around 1,200,000 Irish immigrants entered the United States.[20]

From the start American "nativists" feared the new immigrants and objected to the idea that the Irish should gain American citizenship. In the 1840s a group of nativists in New York founded the American Republican Party, which spread as the Native American Party (its nickname was the "Know Nothing" Party, since its members were instructed to reply "know nothing" when they were asked about the activities of the party). It was successful in the elections of 1854 and 1855, but then fell apart because of internal disagreements. One of the major concerns of the nativists was that Catholics were undermining the Protestant state, leading to an undemocratic situation where Americans would be under the control of the Pope. Anti-Catholicism would remain an issue in the United States up until the election of President John F. Kennedy in 1960.

Between 1880 and 1914, approximately 7,500,000 immigrants from Southern, Eastern and Central Europe entered the United States, including four million Italians. Among the Central and Eastern Europeans was a substantial contingent of about 200,000 Jews. In politics, the arrival of new European groups in the United States led to an effort to limit immigration, starting with the 1894 Immigration Restriction League. These efforts resulted in various immigration restriction laws, such as those of 1917, 1921, and 1924.[21]

The encounters with the new groups of immigrants added to the complexity of American images of Europe. For example, Henry James, who loved Europe so much he finally moved there permanently, noted his ambiguity about Italians in *The American Scene*, a treatise based on

---

19 In recent decades the term "Native Americans" is used as the politically correct terms for the Indian nations.

20 Peter I. Rose, *They and We: Racial and Ethnic Relations in the United States*, 6th ed., Boulder, Paradigm, 2006, pp. 52-53.

21 *Ibid.*, pp. 57, 64, 63. On the immigration restriction laws see Mae M. Ngai, *Impossible Subjects: Illegal Aliens and the Making of Modern America*, Princeton, Princeton UP, 2005.

his visits back to the United States in 1904 and 1905. Unlike the nativitists, James believed that the new immigrants would become American, if not immediately, then by the second generation. Thus he perceived the Italians he met in New York to be different from the Italians he met while in Italy. He explained the difference by referring to the process of Americanisation – which was not necessarily a bad development, as James already was very condescending about Italians apart from the elites and their culture (and even more condescending about African Americans and Chinese). As James wrote about Italian immigrants,

> The Italians, who, over the whole land, strike us, I am afraid, as, after the Negro and the Chinaman, the human value most easily produced, the Italians meet us, at every turn, only to make us ask what has become of that element of the agreeable address in *them* which has, from far back, so enhanced for the stranger the interest and pleasure of a visit to their beautiful country. They shed it utterly, I couldn't but observe, on their advent, after a deep inhalation or two of the clear native air; shed it with a conscientious completeness which leaves one looking for any faint trace of it.[22]

James's complicated image of Italy included a rather primitive, and for that very reason "agreeable," Italian national character, that became less "colourful," as he put it, when Italians became Italian Americans.

It should not be surprising that an Italian American like the movie director Martin Scorsese had a very different image of Europe and, specifically, Italy. He remembered watching Italian movies on an Italian television station in New York with his extended family and friends, just after World War II. His grandparents had emigrated from Sicily but had never told him about their life there. When they watched a movie like *Roma, Città Aperta* about Rome at the end of the war, Scorsese recalled:

> Some of these images were so moving, so intensely moving, that I noticed that my grandparents started to cry. I mean, they left Sicily in 1909, 1910. They made their way all the way to America and here they were, seeing on television in effect, what they had left behind and what had happened to it, what had become of it. When I looked at their faces I could see that this was who my grandparents really were and that this was the country they called home. And it was overwhelming because for the first time I became aware where I came from. I saw them suffering with the people in the film. They felt protected here in America and I could see that they were feeling relief and guilt at the same time. That could have been them up there on the screen.

This sense of having left behind a possibly desperate life became even stronger for Scorsese when he watched another Rossellini movie about the end of World War II in Italy: "In the Naples episode of *Paisà* I

[22] Henry James, *The American Scene*, London, Chapman & Hall, 1907, p. 129.

saw something that really shocked me. I realized that if my family had never left Italy I could have been one of the kids up there on the screen – reduced to stealing shoes, selling them on the black market".[23]

How complicated the images of Europe could become was illustrated by American Studies scholar Daniel Aaron. In the early 1960s he was invited to lecture in Poland. His image of Poland was partly informed by knowing Polish Americans and partly by vague notions about life in Poland, a nation to visit perhaps but not one in which to live:

> My picture of Poland had been compounded from a mix of fact, fiction, and imagination. A large number of my fellow citizens in Northampton, Massachusetts, as I've said, were the children and grandchildren of Polish immigrants, and Galician Polish could be heard on Main Street and in the fields and orchards east of the Connecticut River, where, for several summers, I had suckered tobacco plants, weeded onions, cut asparagus, and harvested apples for Polish-American farmers. But Poland was for me a phantasmagorical map dotted with unpronounceable names, a blood-drenched land in the Eastern fens, comprised of dark forests, fields of sugar beets, burning ghettos, and obscene encampments stacked with human hair and children's shoes. I had chosen Poland over Hawaii (where I had also been invited), because [...] I wanted to know what it was like to live in a Communist society. Warsaw lacked palm trees, but it was said to be au courant. Polish movies were in vogue. Friends returning from brief visits spoke enthusiastically of the Warsaw scene, of sopping up vodka and gossip at the bar of the Bristol Hotel with avant-garde artists and intellectuals.[24]

Aron's image was a somewhat negative but intriguing one, of a "dangerously attractive elsewhere", as he called Europe.[25] His perception of Poland included a notion of a bohemian, artistic life. To participate in that life, Americans needed to go to Europe, and many did.

American Bourgeois and Bohemian Fantasies of Europe

Even though Americans might have seen themselves as being ahead of many European nations when it came to politics, religion, the economy, or social matters, there was also the fixed idea that Europe was culturally superior. Henry James already had stated it clearly in 1879, when he wrote about the United States:

> No state in the European sense of the word, and indeed barely a specific national name. No sovereign, no court, no personal loyalty, no aristocracy, no church, no clergy, no army, no diplomatic service, no country gentle-

23 *My Voyage to Italy*, Miramax Films, 1999. Directed by Martin Scorsese.

24 Daniel Aaron, *The Americanist*, Ann Arbor, University of Michigan Press, 2007, pp. 104-105.

25 *Ibid.*, 92.

men, no palaces, no castles, nor manors, nor country-houses, nor parsonages, nor thatched cottages, nor ivied ruins; no cathedrals, nor abbeys, nor little Norman churches; no great Universities nor public schools – no Oxford, nor Eton, nor Harrow – no literature, no novels, no museums, no pictures, no political society, no sporting class – no Epsom nor Ascot![26]

The Boston art collector and critic James Jackson Jarves agreed with Henry James when he wrote, "The Great Exposition of 1867 at Paris taught us a salutary lesson by placing the average American sculpture and painting in direct comparison with the European, thereby proving our actual mediocrity".[27] Adam Gopnik, the Paris correspondent for *The New Yorker* from 1995 to 2000, indicated that this feeling of inferiority might never be overcome, when he wrote,

I arranged to meet my friend Nicole Wisniak at the Café de Flore, on the boulevard Saint-Germain, for lunch. Nicole is the editor, publisher, advertising account manager, and art director of the magazine *Egoïste* and is a woman of such original chic that in her presence I feel even more ingenuous and American than I usually do, as though pinned to the back of my jacket were a particularly embarrassing American license plate: "Pennsylvania: The Keystone State" or "Explore Minnesota: 10,000 Lakes.[28]

This American sense of European cultural superiority continued, inducing many Americans to travel to Europe, either as tourists or as a "cultural apprentices". Gopnik divided these American visitors into two groups, the bourgeois and the bohemian:

Happiness is multiple, and ours tends to cluster around two poles: one essentially bourgeois, the other bohemian. There are Americans who come to Paris for the food, so to speak. They are in Paris to enjoy the haute-bourgeois civilisation of comfort and pleasure and learning and formal beauty – to walk in the Luxembourg Gardens and study at the Bibliothèque and sketch at the Louvre. In contrast are those who [...] come for the drink – for the dazzlement of new art and new experiences, and, usually, for the drinks, too.[29]

This dichotomy was true for many Americans of artistic temperament. Henry James, although he preferred Rome and especially London

---

26 Henry James quoted in John Lamberton Harper, *American Visions of Europe: Franklin D. Roosevelt, George F. Kennan, and Dean G. Acheson*, New York, Cambridge UP, 1994, p. 8.

27 Carol Troyen, "Innocents Abroad: American Painters at the 1867 Exposition Universelle, Paris", *American Art Journal*, 16/4 (1984), pp. 2-29, quote on p. 14.

28 Adam Gopnik, *Paris to the Moon: a Family in France*, London, Vintage, 2001, pp. 78-79.

29 Adam Gopnik quoted in Adler, "'We'll Always Have Paris'", p. 14.

to Paris, generally travelled for the bourgeois experience in Europe. His novels about Americans in Europe, from *Daisy Miller* or *Portrait of a Lady* to *The Ambassadors*, had as their main theme the mores and behaviour in *haute bourgeois* European circles. As Edith Wharton, who also wrote society novels like James, said about him, "Henry James was essentially a novelist of manners, and the manners he was qualified by nature and situation to observe were those of the little vanishing group of people among he had grown up […]".[30] James also had a romantic view of finding "Art" in Europe. As he recalled of an early visit to Paris,

> The […] quays, with their innumerable old bookshops and print-shops, the long cases of each of these commodities […] must have come to know us almost as well as we knew them; with plot thickening and emotion deepening steadily, however, as we mounted the long, black Rue de Seine – *such* a stretch of perspective, *such* an intensity of tone as it offered in those days; where every low-browed vitrine waylaid us and we moved in a world of which the dark message, expressed in we couldn't have said what sinister way too, might have been "Art, art, art, don't you see? Learn, little gaping pilgrims, what that is!".[31]

And in the Galerie d'Apollon James wrote that he "inhaled little by little, that is again and again, a general sense of *glory*. The glory meant ever so many things at once, not only beauty and art and supreme design, but history and fame and power, the world in fine raised to the richest and noblest expression".[32]

Ernest Hemingway and Henry Miller, like James, were driven to Europe by artistic impulse, but they were much more interested in the bohemian version of an American in Europe. Hemingway celebrated his Parisian life in his memoir *A Moveable Feast*. And Henry Miller wrote, "It is now the fall of my second year in Paris. […] I have no money, no resources, no hopes. I am the happiest man alive. A year ago, six months ago, I thought that I was an artist. I no longer think about it, I *am*".[33]

In spite of original American writers such as Nathaniel Hawthorne, Herman Melville, Emily Dickinson, and Mark Twain, many American authors continued to believe that in order to learn to write one should visit Europe. Some of them made Europe their permanent residence, such as James, T.S. Eliot, Ezra Pound, and Gertrude Stein. Others, like

---

30 Edith Wharton, *A Backward Glance: an Autobiography*, New York, Touchstone, 1998, p. 176.

31 Morton Dauwen Zabel (ed.), *The Portable Henry James*, New York, Viking, 1960, p. 554.

32 *Ibid.*, pp. 560-561.

33 Henry Miller, *Tropic of Cancer*, Paris, Obelisk, 1957, p. 11. See also Ernest Hemingway, *A Moveable Feast*, London, Arrow, 2004.

Ernest Hemingway, F. Scott Fitzgerald, and Henry Miller, stayed in Europe for prolonged periods of time. After World War II, American authors would continue to visit Europe, testifying that even though the United States had become a military and political superpower beyond doubt, Europe remained a cultural h(e)aven. Among such authors were Gore Vidal, Tennessee Williams, Truman Capote, Ralph Ellison, Mary McCarthy, Saul Bellow, James Baldwin and Richard Wright.[34]

For the African Americans, like Ralph Ellison, James Baldwin and Richard Wright, Europe also meant an escape from racism in the United States. Even though discrimination took place in Europe as well, these writers, like the African American soldiers during both World Wars, found that the living conditions in countries such as the United Kingdom or France were more congenial than in the United States. As James Baldwin stated, "I went there [Paris] to get enough away from The American Negro Problem – the everyday insults and humiliation, the continual sadness and the rage – so that I could sit down and write with half a clear head". When he returned to the United States in 1977, however, Baldwin believed that the situation in Europe had changed: "Now many of the former slaves of the Western colonial empires have come to Europe: blacks, Mohammedans, Pakistanis, Moluccans. A Harlem is rising in Paris".[35]

The enthusiasm shown by literary authors for Europe was certainly shared by American painters. In the tradition of the Grand Tour, they visited Europe, often for prolonged periods. Some of them became famous, such as James Abbott McNeill Whistler, Thomas Eakins, John Singer Sargent, and Man Ray. Their experience confirmed something Henry James had written in 1887: "It sounds like a paradox, but it is a simple truth, that when to-day we look for American art, we find it mainly in Paris. When we find it out of Paris, we at least find a great deal of Paris in it".[36] Art historian Barbara Weinberg confirmed James' perception when she wrote that the Hudson River School of landscape painting, celebrating nature in the United States (as an answer to the way in which culture and history were proper subjects for European paintings), was replaced by the influences of American painters who had been to Europe: "[...] many American painters turned increased atten-

---

34 Pells, *Not Like Us*, pp. 6, 139-140.

35 James Baldwin, quoted in Robert Coles, "James Baldwin Back Home", *New York Times Book Review*, 31 July 1977.

36 Henry James quoted in Adler, "'We'll Always Have Paris'", p. 11.

tion to the human figure, largely in response to their exposure to Parisian standards".[37]

Americans knew about art in Europe through various means. The illustrated press, including art magazines, had flourished in the United States since the 1870s. Foreign periodicals were available, but there also existed American journals that paid attention to foreign artists and exhibitions. According to art historian Michael Quick, "Jean François Millet and Mariano Fortuny were names known and discussed in Midwestern towns, where it was possible to follow the main success of the Paris Salon (and the fortunes of the Americans exhibiting in it)".[38] American museums also played an important role in augmenting the appeal of European art and Europe. When Hemingway was in Paris, one way to gain inspiration was to look at Impressionist paintings:

> […] the Cézannes and […] the Manets and the Monets and the other Impressionists that I had first come to know about in the Art Institute at Chicago. I was learning something from the painting of Cézanne that made writing simple true sentences far from enough to make the stories have the dimensions that I was trying to put in them.[39]

For American women painters, too, being in Europe was attractive – and for additional reasons as well. As art historian Erica E. Hirshler pointed out, a stay in Europe meant that these women gained an autonomy which they did not have in the United States. In addition, they did not even have to follow the same social rules as French women. Hirshler quotes the American painter May Alcott, who lived and worked with her family near Paris, and had said, "in America this cannot be done, but foreign life is so simple and free, we can live for our own comfort not for company".[40]

If for some American women Europe was seen as an escape from bourgeois restrictions that made it difficult to combine family life and an

---

37 H. Barbara Weinberg, "Summers in the Country", in Adler *et al.*, *Americans in Paris*, p. 150.

38 Adler, "'We'll Always Have Paris'", p. 27.

39 Hemingway, *A Moveable Feast*, p. 8.

40 Erica E. Hirshler, "At Home in Paris" in Adler *et al.*, *Americans in Paris*, p. 84. May Alcott was the younger sister of Louisa May Alcott, the author of *Little Women*. In the novel, Louisa May expressed the interest in Europe when she had one of her characters, Jo, cry out, "Abroad! Oh, tell me about it! I love dearly to hear people describe their travels", and then say, "Don't I wish I'd been there! Did you go to Paris?". In the second part of *Little Women*, Amy (the literary version of May) travelled to Europe. In 1879, after the tremendous success of *Little Women*, Louisa began writing a novel about two women artists, *Diana and Persis*, based on her sister's life in Europe. She stopped working on the novel when May died in childbirth that same year (Louisa May Alcott, *Little Women*, New York, Penguin, 1989, pp. xx-xxi, 28).

artistic existence, for other Americans, women and men, Europe was an abandonment of the restrictions on their romantic and sexual lives. Henry James described such a woman in his 1878 novel *Daisy Miller*, whose main character freely consorted with men, without a chaperone, to the extent that her compatriots in Europe shunned her. In the 1930s, and in more explicit ways, Henry Miller made his sexual pursuits part of his bohemian lifestyle. Similarly, in James Baldwin's *Giovanni's Room*, written in 1958, the American David goes to France "to find himself", ultimately to admit to his homosexuality after all. In 1957, Elaine Dundy had her Sally Jay Gorce in *The Dud Avocado* explain the source of her adventurous love life in Paris in these terms: "because I feel the need to sin".[41]

In the years after the end of the Second World War, the bourgeois and bohemian ideals of life in Europe began to spread beyond the wealthy, the writers and the artists to the American middle classes as well. From 1958 onward, commercial transatlantic flights with jet airplanes made it possible for middle-class Americans to travel to Europe on short trips.[42] Their image of Europe was informed not only by travel literature and high art but also by popular culture. When Adam Gopnik and his family decided to go to Paris, in their case for a five-year stay during the 1990s, he recalled the influence of his earlier images of Paris: as a child, having "a three-dimensional cutout of a Parisian policeman" in his bedroom, "the book adaptation of *The Red Balloon*, the wonderful Albert Lemorisse movie about a small boy in the Parisian neighborhood of Menilmontant", "the overbright streets of the Madeline books, covered with vines and the little girls neat in their rows, and black and white pictures of men in suits walking through the Palais Royal, takes from a Cartier-Bresson book on the coffee table". As a teenager he made two visits to Paris and watched French movies starring Jacqueline Bisset; and as an adult, he made more visits to Paris, saw the movie *The Umbrellas of Cherbourg*, and "listen[ed] to old Charles Trenet records".[43]

For the Americans who flew to Europe in the 1950s and later, and specifically to visit Paris, there were other images, too. For instance, there were the fashion reports from Paris in American magazines, showing how elegant, chic, and sophisticated Paris was. From the 1940s until the early 1960s, Richard Avedon, for one, had shot photo series in Paris for *Harper's Bazaar* magazine. They were staged events in which

---

41 Elaine Dundy, *The Dud Avocado*, London, Virago, 2001 (first published 1958), p. 152.

42 Pells, *Not Like Us*, p. 137.

43 Gopnik, *Paris to the Moon*, pp. 5-9.

the models were like actors in a drama. In 1956 and 1957, Avedon made photo series at famous spots such as Place de la Concorde, Place Vendôme, and backstage at the Folies-Bergère. Probably the most well-known picture of this series was of Dovima with elephants at the Cirque d'Hiver, shot in 1955.[44] In 1962, Avedon created, with models Suzy Parker and Mike Nichols, a bogus news story for *Harper's Bazaar*. The story was in the style of paparazzi photographers and was a take on the attention surrounding the famous spats between Elizabeth Taylor and Richard Burton who were filming the movie *Cleopatra*.[45]

Richard Avedon's photo shoots themselves became the inspiration for a Hollywood movie, *Funny Face*, with Fred Astaire playing the role of photographer and Audrey Hepburn as a bookseller being turned into fashion model. The movie basically used Paris as a background, while Astaire danced, and he and Hepburn sang George and Ira Gershwin songs. In a true blending of the bourgeois and the bohemian, fashion was used to convey bourgeois chic, while the character Hepburn played is drawn to "emphaticalism" (the movie's simplistic version of existentialism) – until the French professor teaching the topic tries to seduce her, making her realise she really loves the (American) photographer (it is a Hollywood movie, after all).[46]

Although this was neither the first nor best Hollywood movie about Americans in Europe, it did contain a number of themes that occurred regularly in the genre. Audrey Hepburn made seven of this type of movie, six of which took place in Paris.[47] While other American films used Europe as a backdrop, I shall focus here on movies in which the American characters actually comment on being in Europe.[48]

---

44 Richard Avedon, *Evidence, 1944-1994*, New York, Random House, 1994, pp. 48-51, 53.

45 *Ibid.*, pp. 71-75.

46 *Funny Face*, Paramount Pictures, 1957. Directed by Stanley Donen.

47 The seven movies are: *Roman Holiday* (1953), *Sabrina* (1954), *Funny Face* (1957), *Love in the Afternoon* (1957), *Paris When It Sizzles* (1963), and *Charade* (1963), *How to Steal a Million* (1966).

48 Hollywood movies in which Americans visit or live in Europe, using the European setting as a background, but without any interesting comments, include *Gentlemen Prefer Blondes*, Twentieth Century Fox, 1953, directed by Howard Hawks (Jane Russell and Marilyn Monroe go to Paris for romance, rich men, and to enjoy their shopping); *To Catch a Thief*, Paramount Pictures, 1955, directed by Alfred Hitchcock (a Hitchcock movie on the French Riviera); *Paris When It Sizzles*, Paramount Pictures, 1963, directed by Richard Quine (Paris as a setting for the writing of a scenario); *Charade*, Universal, 1963, directed by Stanley Donen (a thriller in Paris); *Cabaret*, Allied Artists Pictures Corporation, 1972, directed by Bob Fosse (1930s Berlin); *The Talented Mr. Ripley*, Miramax Films/Paramount Pictures, 2000, directed by Anthony Minghella (even if the movie takes place in Italy, there is

Hepburn's first Hollywood movie was *Roman Holiday* (1953), in which the image of romantic love and bohemian life was depicted through the story of a European princess (of an unknown country) fleeing her courtiers and obligations to enjoy life in Rome with an American journalist (played by Gregory Peck; another, and rather strange theme of the Hepburn movies was that her love interests were always men at least twenty years her senior). A link with the earlier topic of American images of politics in Europe is that the first question the journalist asks, at the behest of his editor, is: "How does Her Highness react to the idea of a European Federation?".[49]

Romance was a crucial theme in these movies: Europe was seen as almost intrinsically romantic. In *Three Coins in the Fountain* (1954) three American secretaries living in Rome are looking for love. They find it after each throws a coin in the Trevi Fountain.[50] Another example is *Love in the Afternoon* (1957), a Audrey Hepburn movie in which she plays a young woman who seduces a millionaire American playboy (Gary Cooper).[51] In *Come September* (1961), Rock Hudson plays an American businessman who owns a villa in Italy, where he stays every September with his Italian girlfriend (Gina Lollobrigida). When they arrive early, they find the villa turned into a hotel by the housekeepers, and rooms rented out to a group of American girls, pursued by a group of American boys. The involvement with these romantic escapades leads to a marriage between the American businessman and his Italian girlfriend.[52] A more recent example is *French Kiss* (1997), in which Meg

---

little interaction with Italian society. There is a scene in which the main characters sing "You wanna be Americano" in Italian, in a jazz club. The song is originally from the movie *It Happened in Naples* (see below), in which Sophia Loren plays an Italian nightclub singer, and "You wanna be Americano" is one of her trademark songs); and *Ocean's Twelve*, Warner Brothers, 2004, directed by Steven Soderbergh (Amsterdam, Paris, Rome).

49 *Roman Holiday*, Paramount Pictures, 1953. Directed by William Wyler. The script was written by Dalton Trumbo (better known for his script for *Spartacus*), who was blacklisted in Hollywood for not being willing to testify in the hearings of Senator Joseph McCarthy on Communism in Hollywood. Perhaps one way to explain why the princess in the movie is shown to be just an ordinary person, although a bit otherworldly, is by pointing to Trumbo's leftist sympathies. In any case, Trumbo created an original movie character (posthumously receiving an Oscar for the script in 1993), and Audrey Hepburn received an Oscar for her role as the princess.

50 *Three Coins in the Fountain*, Twentieth Century Fox, 1954. Directed by Jean Negulesco. The use of the Trevi Fountain in a romantic context is very different from the way Federico Fellini would use it six years later in *La Dolce Vita*, when he had Anita Ekberg and Marcello Mastroianni frolic around in it.

51 *Love in the Afternoon*, Warner Brothers, 1957. Directed by Billy Wilder.

52 *Come September*, Universal International Pictures, 1961. Directed by Robert Mulligan.

Ryan as an American tries to win back her fiancée (who had fallen in love with a French woman). She is aided by a French petty thief (Kevin Kline), whom she helps to get on the straight path and with whom she eventually falls in love.[53] Probably the ultimate romantic movies in this context are *Before Sunset* (1995) and *Before Sunrise* (2004), in which an American tourist (Ethan Hawke) and a French woman (Julie Delpy) meet each other for brief encounters, in Vienna and Paris, and are twice confronted with the question of whether they care enough about each other to change their lives and stay together.[54]

Royalty was another theme of Hollywood movies: Americans were fascinated by an institution they did not have but which represented, perhaps, something for which they feel a strange nostalgia. *Royal Wedding* (1951) took place with the wedding of Princess Elizabeth and Prince Philip as a background to the story.[55] *A Breath of Scandal* (1960) played at the Imperial court in Vienna.[56] In a later movie, *National Lampoon European Vacation* (1985), the main character dreams of meeting (by then) Queen Elizabeth.[57]

Perhaps above all, Europe was glamorous to American audiences. In *Sabrina* (1954), Audrey Hepburn played a young woman in whom men were not interested, until she had lived in Paris and learned Parisian ways. As she wrote to her father about her return to Long Island: "If you should have any difficulty recognising your daughter, I shall be the most sophisticated woman at the Glen Cove station".[58] *An American in Paris* (1951) stresses the glamorous life as well, both from the bourgeois perspective of a rich American heiress and the bohemian perspective of an American painter (Gene Kelley) – poverty has never been so much fun as in this movie, with songs by George and Ira Gershwin.[59]

The columnist Art Buchwald would refer to these movie images in his memoirs, while writing about his arrival in Paris, adding a touch of reality:

---

53 *French Kiss*, Metro Goldwyn Mayer, 1997. Directed by Lawrence Kasdan.

54 *Before Sunrise*, Castle Rock Entertainment, 1995. *Before Sunset*, Warner Brothers, 2004. Directed by Richard Linklater.

55 *Royal Wedding*, Metro-Goldwyn-Mayer, 1951. Directed by Stanley Donen. The movie is renowned for the scene in which Fred Astaire dances on the walls and ceiling of a room.

56 *A Breath of Scandal*, Paramount Pictures, 1960. Directed by Michael Curtiz. Based on a play written by Ferenc Molnar, adapted by Sidney Howard.

57 *National Lampoon European Vacation*, Warner Brothers, 1985. Directed by Amy Heckerling.

58 *Sabrina*, Paramount Pictures, 1954. Directed by Billy Wilder.

59 *An American in Paris*, Warner Brothers, 1951. Directed by Vincente Minelli.

> As I walked out of the Gare St-Lazare at about nine o'clock in the evening that day, my first impression of Paris was that it wasn't so much a city as a stage setting from a Broadway musical. The sidewalk cafés were exactly as they were depicted in the magazines and movies. There were honking taxis, street vendors selling dirty pictures, roses, oriental rugs, black-market money-changers from Corsica, men peddling carved-ivory ornaments from who-knows-where, begging gypsies with babies on their shoulders, and, finally, heavily rouged streetwalkers willing to sell sexual gratification for a price.[60]

Even though Buchwald, choosing a bohemian lifestyle that he could afford on his GI Bill income, noted the harsher side of life in Paris, it was nevertheless the movie images that formed his original ideas of the City of Light. In any case, Buchwald enjoyed his life in Paris, warts and all.

Of course, all was not romance, glamour, and happy endings in American films set in Europe. Another major theme in Hollywood movies was the lack of morals in Europe. One way to make a moral statement was by putting an American businessman in a "typical" European setting. In *A Breath of Scandal* (1960) it was the practical, bourgeois businessman at the pre-World War I Austrian court. He tries to marry a princess (played by Sophia Loren), while she is being married off to a Prussian prince. In the end she chooses the American businessman, but only after her father tells her he had married out of love too. In a bow to European duplicity, he pretends to have a mistress, "just to preserve my social standing", and he keeps his marriage out of love a family secret, lest "people would think we were undermining the government".[61]

In *It Started in Naples* (1960), an American businessman (Clark Gable) goes to Naples to settle the estate of his brother. When he roams around the streets after midnight, because loud music played for the tourists prevents him from sleeping, he asks a waiter "Does this go on all night? You know, it's after 1:00. How are people supposed to sleep on this island?". The waiter replies, "Together".[62]

In *Avanti!* (1972) another American businessman (Jack Lemmon) becomes upset when he learns that lunch takes three hours in Italy. He is told by a local "here we do not rush to drugstore for chicken sandwich and Coca-Cola. Here we take our time. We cook our pasta. We sprinkle

---

60 Art Buchwald, *I'll Always Have Paris! A Memoir*, New York, Fawcett Columbine, 1996, pp. 4-5.

61 *A Breath of Scandal*, Paramount Pictures, 1960. Directed by Michael Curtiz.

62 *It Started in Naples*, Capri Pictures/Paramount Pictures, 1960. Directed by Melville Shavelson, scenes 5 and 6.

our *parmigiano*. We drink our wine. We make our love". When the businessman asks: "What do you do in the evening?", the answer is: "In the evening we go home to our wives".

In Hollywood movies made more recently, the references to the decadence of Europe become increasingly explicit. *National Lampoon European Vacation* (1985) seizes the opportunity to show women topless (in a Paris nightclub and at a German *bierfest*). In movies apparently made for a teenage audience, like *Deuce Bigelow: European Gigolo* (2005) and *Eurotrip* (2006), the scenes shown includes visiting a bar ("there is no drinking age"), soccer hooligans, a nude beach, and visits to Amsterdam ("the sex and drugs capital of Europe" or "Disneyland for college students") with images of drug use and of the Red Light district.[63] In *Eurotrip* one of the main characters summarises his image of, and plans for, Europe as follows:

> I am talking about crazy European sex. You know America was founded by prudes. Prudes who left Europe, because they hated all the kinky, steamy European sex that was going on. And now I, Cooper Harris, will return to the land of my perverted forefathers and claim my birthright, which is a series of erotic and sexually challenging adventures.[64]

This last remark could also be taken as a comment on United States culture. A good example of a movie criticising both cultures was the comedy *One, Two, Three* (1961) by director and immigrant Billy Wilder. The movie was shot in Berlin. The opening shot notes in writing that the story takes place just before the Berlin Wall was built, a time of high international tension. The story is about the president of the local Coca Cola plant, C.R. MacNamara, played by James Cagney, who is trying to get a promotion by setting up a deal with the Soviets. He has to "babysit" the daughter of his boss, who is on a tour of Europe.

First, Wilder comments on American naïveté, as in the scene when the young woman, Scarlett Hazeltine, arrives and MacNamara prevents her from making a date with a French airline steward:

> Scarlett Hazeltine: "So you're gonna be like that?"
>
> MacNamara: "Like what?"
>
> Scarlett: "A company man like those old poops in Rome and Paris breathing down my neck every minute, cramping my style."
>
> MacNamara: "Look, Miss Hazeltine. You're underage, and I'm under orders."

63 *Deuce Bigelow: European Gigolo*, Columbia Pictures, 2005. Directed by Mike Bigelow. *Eurotrip*, Dreamworks, 2006. Directed by Jeff Schaffer.

64 *Eurotrip*, scene 5.

> Scarlett: "Europe – what a drag! I've done the Colosseum bit and the Mona Lisa bit but they never take me to any of those marvy places like the Lido and Crazy Horse and Le Sexy."
>
> MacNamara: "I promise you, you'll have a wonderful time in Berlin."
>
> Scarlett: "That's why I came. I hear this is a real swinging town."
>
> Mrs MacNamara: "Where did you hear that?"
>
> Scarlett: "Don't you ever read the headlines? Everybody says Berlin's the hottest spot in the world right now."

Second, Wilder uses European politics as a way to comment on American feelings of superiority in the 1960s. Scarlett gains a husband from East Berlin, whom she helps blow up balloons with the text "Yankee Go Home". MacNamara worries: "You've been helping this guy spread anti-American propaganda?" To which Scarlett, born and raised in the American South, replies: "It's not anti-American. It's anti-Yankee. Where I come from, everybody's against the Yankees".[65]

Conclusion

Billy Wilder's observations on American images of Europe return us to the ideas of Mark Twain and the painter William Merritt Chase, quoted at the beginning of this chapter. American images of Europe often are comments on American society. A remark like "I would rather go to Europe than go to Heaven" not only ranks Europe higher than heaven, but definitely ranks the United States below either option. Billy Wilder's observation on the legacy of the Civil War in the United States pointed out to Americans that feeling (politically) superior to Europe in the early 1960s was misplaced.

American images of Europe are also exactly what one would expect: American observations of European countries and people. In this chapter I have analysed a number of themes among the American images of Europe. In politics, there is the persistent distrust of Europe as the origin of international conflicts in which the United States might become embroiled. Immigration also led to negative American images of Europe. Europeans moved to the United States for economic, political and/or religious reasons. Yet, when it was about culture, Americans admired Europe, and wanted to go there, to visit, to learn, or even to stay permanently.

The key point to be taken is that these themes should be considered in a context of a "multiplicity of notions". It really depends which American makes observations, on what particular part of Europe, and

---

65 *One, Two, Three*, The Mischkin Company, 1961. Directed by Billy Wilder.

during which time period. Wealthy, nineteenth-century New Yorker Henry James had a different image of Italians from the twentieth-century Italian American New Yorker Martin Scorsese, raised in Little Italy. A Jewish American, like Daniel Aaron, probably had a different view of Poland from that of many Polish Americans. Positive images of Europe had often negative counterparts. Where some Americans were looking for the life of a bohemian in Europe, other American saw moral depravity. In politics, Americans have not always thought of their former coloniser Britain in the sense of a "special relationship", or of France, the country of La Fayette, as an opposing force in the world. In short, there is a wide range of American images of Europe, and they are used depending on the situation, political or otherwise, whether in the United States itself or in American-European relations.

CHAPTER 7

Encountering the European and Western Other in Chinese Propaganda Posters

Stefan R. LANDSBERGER

After the People's Republic of China (PRC) opened its doors to the outside world in the late 1970s, living conditions improved. Owning a radio or even a television came within everyone's reach, more than ever before. Print publications ranging from newspapers to journals and magazines also mushroomed in an unprecedented way. Increasingly driven by the demands of a burgeoning market economy, the media were forced to depoliticise their content and to become more responsive to the needs of their audience.[1] As a result, the central state and party apparatuses lost their monopoly over ever larger number of channels of dissemination and interpretation of information. More and more, the static produced by non-political and polysemic media crowded out the singular, authoritative Voice of China that was once provided by the Chinese Communist Party (CCP).

These seemingly unrelated developments nonetheless were to have a lasting effect on the way the Chinese interacted with the outside world. One result was that more images of events and peoples outside China started flooding in, and they could no longer be framed by the government in the same manner as before. This framing had previously had an enormous impact on the way in which these "foreigners" had been depicted and their images had been presented. In many ways, the assigning of standardised, predetermined complexes of interpretation to visual information involved in this process resembled the Chinese penchant for formalising language into *tifa*, formulations or set phrases, preferably including numbers ("Three-Eight Work Style", "One Country, Two Systems", etc.).[2] The repetitive use of these visual *tifa*, then, greatly

1 Chin-Chuan Lee, Zhou He, & Yu Huang, "'Chinese Party Publicity Inc.' Conglomerated: the Case of the Shenzhen Press Group", *Media, Culture & Society*, 28/4, 2006, p. 583.

2 Michael Schoenhals, *Doing Things with Words in Chinese Politics*, Berkeley, Institute of East Asian Studies, 1992, pp. 2-3, 6.

influenced the way the Chinese population had seen and interpreted the world in the recent past.

When analysing the way in which the Chinese visualised Europeans through time, we are confronted with one major problem: the Chinese do not differentiate between French and Americans, between Germans and Canadians. To them, all whites, or Caucasians, are more or less the same. This has less to do with the Chinese penchant for formalised representation, and more with Chinese ideas about race, as will become clear below. In order to discuss the changes in the way Europeans were represented in Chinese imagery, then, we are obliged to look at the representation of all Caucasians.

In the three decades following the founding of the People's Republic in 1949, the appearance of the Other in the media had been carefully crafted and determined largely by China's identification of friends and foes in the international community. Propaganda posters played an important role in the construction of the image of the non-Chinese Other, supplementing other visual information provided by *Xinhua* (New China News Agency) newsreels and press photographs. Most of these images basically served a function within the framework of Chinese Occidentalism, the "deeply rooted practice of alluding to the Occident as a contrasting Other in order to define whatever one believes to be distinctively 'Chinese'",[3] and helped the young nation to demarcate its proper place in the world.[4] The visual information was intended to demonstrate the legitimacy and instrumentality of CCP rule, on the one hand by showing friends and allies, on the other hand by providing enemies posing a threat to the nation. At the same time, they served as examples proving that *The Thoughts of Chairman Mao*, the ideology that underpinned the Chinese nation, was the supreme principle and living gospel of both Marxism and Leninism, and that its application elsewhere had the same beneficial effects.[5] As domestic politics radicalised from the late 1950s onwards, China found itself left with a steadily dwindling number of allies abroad. The endless stream of both artistic images and photographs released by *Xinhua* showing foreigners visiting China and meeting with the leadership were intended to demonstrate that China indeed still mattered. They were an important mechanism to prove that China continued to have friends all over the world, even when the country seemed to be more isolated than ever.

---

3 Xiaomei Chen, *Occidentalism – a Theory of Counter-Discourse in Post-Mao China*, New York, Oxford UP, 1995, p. 39.

4 Peter Hayes Gries, *China's New Nationalism – Pride, Politics, and Diplomacy*, Berkeley, University of California Press, 2004, p. 36.

5 Chen, *Occidentalism*, p. 6.

Although they were produced for a Chinese audience (the slogans, after all, are in Chinese script), it is not clear whether the propaganda posters devoted to these contacts with foreign peoples strongly influenced the popular political consciousness, other than by maintaining and further strengthening the CCP's political hegemony.[6] Nonetheless, I will concentrate on those visual materials that have a strong political tone rather than materials devoted to subjects such as international sports events in which China participated.[7]

The Other Visualised

What can be said in general about the Chinese images of the Western Other? In imperial times, the Chinese considered foreigners to be "barbarians", people existing at the margins of a hierarchical world system in which China, as the most highly developed culture within that system, sat at the centre.[8] The inferior barbarians were strange creatures, with strange ways of dress and even stranger habits, and functioned as sources of amazement and amusement. By the end of the nineteenth century, however, the Chinese became increasingly aware that their feelings of superiority and aloofness from the world were misplaced and were at the root of the sorry state the nation found itself in. Having missed out on the Industrial Revolution and ensuing Imperialism, China had fallen victim to the West's insatiable hunger for markets, resources and influence. Consequently, national attitudes to the foreign presence and interference in Chinese affairs hardened in the first decades of the twentieth century. Once Mao Zedong declared in 1949 that the Chinese people had stood up and would no longer be exploited or victimised by outsiders, one might have expected that most if not all foreigners would disappear from official Chinese imagery, with the exception of convenient outsiders who could function as targets for national anger if things went wrong. Both the United States and Europe neatly fitted that latter category as the Cold War developed in the 1950s.

However, China had decided that in order to resist further Western encroachment, acquiring Western knowledge was unavoidable, and modernisation could be learned best at the knees of the Soviet Union.

---

6 Chen, *Occidentalism*, pp. 15-16.

7 The posters discussed in this chapter are part of two collections, viz., the IISH Stefan R. Landsberger Collection and the IISH Chinese Propaganda Poster Collection, unless otherwise indicated. Parts of these collections can be viewed at http:// www.iisg.nl/~landsberger and http://www.iisg.nl/exhibitions/chairman/ chnintro.php, respectively.

8 Chen Zhimin, "Nationalism, Internationalism and Chinese Foreign Policy", *Journal of Contemporary China*, 14/42, Feb. 2005, pp. 36-37.

China thereby adopted the symbolically significant, subservient persona of the student. As a result, during the period of socialist reconstruction in the 1950s, Soviet experts were represented in official imagery as the "elder brothers" who taught the Chinese what this modernisation was all about. After the Chinese and the Soviets fell out in the late 1950s, however, favorable images of foreigners almost disappeared, until the mid 1960s, when China started to give its support – mostly verbal – to revolutionary and/or liberation movements in Asia, Africa and Latin America. By the early 1970s, China had only a few friends left in the world, and that was echoed in its visual propaganda. Less than a decade later, the Chinese attitude towards the outside world had changed completely again. Co-operation with foreign nations, in particular those from the developed West, was to play a major role in China's development. As a result, and in line with the general depoliticisation of society that got under way in the 1980s, contacts with foreign nationals were visualised in utilitarian rather than political terms. In the process of ongoing integration in the international political and economic order, the Western world ideally no longer functioned as a real or imagined enemy, but as an endless source of information about development that provided examples which could be emulated or rejected.

However, as China's economic achievements since then have done much in assuaging popular feelings of inferiority, they have reawakened a deep-seated nationalistic pride in the nation among the people. The idea that a re-emerging China must take on the mission of modernising and civilising the world, while still coupled with a lingering historic resentment about having been exploited by the West, has resulted in insecurity that once more needs Western countries as a foil to rally against.[9]

Race

When discussing non-Chinese and analysing their images in Chinese media, one cannot avoid the topic of racial prejudice, or better still, the Chinese feelings of supremacy over other races, which evolved from cultural superiority to include an ethnic and even racial component.[10] Of course, China itself claims it knows no racism, as this is seen as a purely Western concept.[11]

---

9 Pál Nyíri, "The Yellow Man's Burden: Chinese Migrants on a Civilizing Mission", *The China Journal*, 56, July 2006, p. 106.

10 Chen, "Nationalism, Internationalism and Chinese Foreign Policy", p. 39.

11 Frank Dikötter, "Introduction", in *idem* (ed.), *The Construction of Racial Identities in China and Japan – Historical and Contemporary Perspectives*, London, Hurst, 1997, p. 2.

From the mid-nineteenth century onward, Chinese reformers were working actively towards the creation of a new symbolic environment intended to replace the inward-looking Confucianism that had prevailed for centuries. The goal of this endeavor was to find a place for China and its people in the bewildering world it saw itself confronted with, a world in which the Sinocentric ordering had been rent asunder. Part of this activity was an interest in other peoples in the world. It resulted in the Chinese division of mankind into a racial hierarchy of biological groups, in which the yellow and white races were seen as being superior to the darker races of brown, black and red.[12] This ordering of mankind positioned the whites (including the Chinese, with their superior cultural achievements) at the centre of the civilised world, whereas black and other coloured peoples were seen as the negative pole of humanity. The latter were relegated to the edge of the known world.[13] The colour black, moreover, was infused traditionally with negative connotations in the Chinese language, as well as in expressions of popular culture such as operas. Even in racy cosmopolitan Shanghai of the 1920s and 1930s, blacks were seen as stereotypical savages.[14] The West, on the other hand, the part of the world where the white race lived, could be respected. It had access to advanced science and technology, and other useful knowledge that the Chinese felt in desperate need of, then as much as today.[15] It would only be a matter of time before the "almost-white" Chinese would regain the superior position they had been accustomed to in the past.

Only after the founding of the PRC were racial relations and problems phrased in terms of class antagonisms. Having suffered the fate of being a semi-colony itself, China was increasingly successful in presenting itself as the leader of all victimised "coloured" peoples in the world in their historical struggle against "white imperialism".[16] In other words, peoples came to be identified as friends or foes on the basis of their relations with white capitalism, and not only through the colour of their skins. The way they were to be treated became an element that was decided upon in the domains of ideology and Chinese foreign policy. An

---

[12] Frank Dikötter, "Racial Discourse in China – Continuities and Permutations", *ibid.*, p. 16.

[13] Dikötter, "Racial Discourse", pp. 32-33; Frank Dikötter, *The Discourse of Race in Modern China*, London, Hurst, 1991, pp. 2-13; Michael J. Sullivan, "The 1988-89 Nanjing Anti-African Protests: Racial Nationalism or National Racism?", *The China Quarterly*, 138, June 1994, p. 441.

[14] William Schaeffer, "Shanghai Savage", *Positions*, 11/1, 2003, pp. 91-133.

[15] Philip Snow, *The Star Raft – China's Encounter with Africa*, London, Weidenfeld & Nicolson, 1988, pp. 189, 206.

[16] Dikötter, *Discourse*, pp. 191-193.

extensive review of the shifts in these domains lies beyond the scope of this paper. Let it be said that relations with non-Chinese nations and nationals continue to be affected by racial considerations, a situation that has not yet changed, to the present day.

White Peoples

Early depictions of representatives of the white race (nationals of European countries and the USA, and the white inhabitants of the colonies of these countries) can be found in the *Dianshizhai Pictorial*, a magazine published in Shanghai that was devoted to current affairs and sensational events at home and abroad.[17] Printed in one of the first stone-based lithographic printing shops in China, it first appeared in 1884 as an illustrated supplement to the Chinese-language newspaper *Shenbao* with the intention of providing pictorial coverage of the Sino-French War which took place in that same year.[18] The *Pictorial* continued independent publication until 1898.

Figure 7.1. "Cricket in Shanghai" (1891)

After Ye Xiaoqing, *The* Dianshizhai *Pictorial – Shanghai Urban Life 1884-1898*, Ann Arbor, Center for Chinese Studies, University of Michigan, 2003.

---

17 Ellen Johnston Laing, *Selling Happiness – Calendar Posters and Visual Culture in Early-Twentieth-Century Shanghai*, Honolulu, University of Hawai'i Press, 2004, pp. 51-53.

18 Christopher A. Reed, "Re/Collecting the Sources: Shanghai's *Dianshizhai Pictorial* and Its Place in Historical Memories, 1884-1949", *Modern Chinese Literature and Culture*, 12/2, 2000, pp. 54, 56.

In all, more than 4,500 journalistic illustrations appeared in the *Pictorial*. Many of the published items featured French, British, American and German nationals sojourning in the city's International Settlement, although Italians and Swedes also were represented.[19] Due to their work as police officers, Sikhs often made an appearance, but they were not considered white by the Chinese readers.[20] Some scenes were located in other Chinese treaty ports where Western foreigners abounded.

The articles (and images) related to Westerners by and large touched upon the following five subjects:[21]

1. The Western community living in China. This group of articles was devoted mainly to customs such as burials and weddings, but also the use of Western cutlery. Other topics were sports events (horse racing, cricket) and national festivities, and *faits divers* in which Westerners occupied a central place (see Figure 7.1);
2. Treaty ports and extraterritoriality. The Westerner was shown as an administrator, maintainer of law and order, and judge. The main body of these reports was situated in Shanghai. It was under this heading that the Sikh police officers appeared. Attention was devoted to the Shanghai Volunteers Corps, and the work of fire brigades using Western techniques to extinguish conflagrations. Missionaries and their activities were other recurring topics;
3. The introduction and application of Western knowledge and technology. These reports dealt with the introduction of railways, medicine, education for women, and the like;
4. The relations between Chinese and Westerners. These reports were tied in with current events. As a result, they revolved around armed conflicts, such as the Sino-French War (1883-85), and the ensuing negotiations between foreign nations and China. The generally negative attitude of Westerners towards the Chinese was regularly touched upon, as were the Chinese feelings towards Westerners. The tone of such reports was both patriotic and sensational;
5. The Westerner as an accidental passer-by in a news item, or Westerners as part of the background. This was often done to lend an aura of authenticity to the report (and the image).

19 Ye Xiaoqing, *The Dianshizhai Pictorial – Shanghai Urban Life 1884-1898*, Ann Arbor, University of Michigan Center for Chinese Studies, 2003, pp. 117-131.

20 Ye, *The Dianshizhai Pictorial*, pp. 1-2, 98-99.

21 This section draws heavily on Rudi Voorrecht, 'Westerlingen in China – een onderzoek naar de berichtgeving in de Dianshizhai huabao 1884-1886, 1896-1898' [Westerners in China – an Investigation of the Reporting in the Dianshizhai Huabao], unpublished MA thesis, Leiden University, 1987, pp. 46-67.

The Westerners in these images were drawn realistically, with great attention to detail. No attempts were made to caricature them, although some of the artists had problems with the Western (round) eyes. Men were depicted with moustaches or long beards, wearing a hat and carrying a stick. Women were shown in the fashion of the day, with curly hair, usually wearing hats, with a sunscreen in their hands. Children were also shown dressed according to the fashion of the day.

The *Dianshizhai huabao*, in short, can be considered as an imaginary stage where Chinese could come into contact with Westerners and their customs, often vicariously. But as the years passed by, and more Westerners ventured into larger parts of China, the experiences were shaped more by the increasing number of "personal" encounters the Chinese had with them. If their attitude had been coloured initially by curiosity, awe and a certain amount of admiration, it became more negative in the late Qing and early Republican period. This was the result of ever-stronger feelings of having been exploited and treated unfairly. The refusal of the Allies during the negotiations over the Versailles Treaty in 1919 to honour the Chinese demands that the Shandong area – formerly under German control – be returned, instead of handing it over to the Japanese, was the turning point in the positive and often admiring attitude towards the West. By the 1930s, the unshaven Western capitalist, drunken, debauched, brutish and loathsome, crouching among moneybags and armaments, or pawing helpless Chinese women, had become a popular stereotype in Chinese progressive caricature.[22]

Friends …

After 1949, two distinct kinds of white peoples were identified: friends and foes. The friendly section was made up of three groups. First and foremost were the Soviets, the "elder brother" (*laodage*) to whose side China had decided to lean in order to learn everything about modernisation. In doing so, China had entered into this bilateral alliance as an explicitly junior partner, thereby making it an essentially unequal relationship.[23] Joining the Soviets, but on a lower level of importance, were their European allies grouped together in the socialist bloc, including East Germans, Poles and Rumanians. For these peoples, a deep proletarian kinship was felt – on the official level at least. The last and least significant group of white friends consisted of the high-profile individual sympathisers that could be exploited for propaganda purposes. They ranged from the American journalist Edgar Snow – whose

---

22 Dikötter, *Discourse*, p. 142.

23 Elez Biberaj, *Albania and China – A Study of an Unequal Alliance*, Boulder, Westview Press, 1986, p. 1.

CCP-sympathetic 1937 book *Red Star Over China* had made Mao a household name in the West[24] – to the French philosopher Jean-Paul Sartre, his wife Simone de Beauvoir and many other representatives of the so-called progressive forces. A special subsection within this group of Western sympathisers was formed by people like Sidney Rittenberg, Israel Epstein, David and Isobel Crook and others, who threw in their lot with the CCP government and, having become "honorary Chinese", took up permanent residence in China as "foreign experts", sometimes even actively participating in the political events of the day.[25] It should be noted that the propaganda objectives served by most of these Western friends were directed largely at the West and were rarely intended for domestic purposes: they were invited to help "explain" abroad how great things were in China.

In the early 1950s, then, images of Russian and East-European politicians, advisers, technicians, workers and scientists of both genders abounded in the Chinese media, including the posters.[26] The tone of these pictorial encounters seems to have been defined by the materials that showed Mao Zedong together with Joseph Stalin. In these images, obvious artists' impressions of Mao's visit to Moscow in 1949-1950, Mao and Stalin invariably stood on the same level (Figure 7.2). This representation, however, was in direct contradiction with the actual relationship between the two countries. Pictorial examples, "templates", of Soviet and other socialist soldiers and civilians were offered to enable all art workers to represent correctly – and uniformly – the new foreign friends in their works.[27]

---

24 Colin Mackerras, *Western Images of China*, Hong Kong, Oxford UP, 1989, pp. 79-80.

25 Anne-Marie Brady, "Red and Expert: China's 'Foreign Friends' in the Great Proletarian Cultural Revolution, 1966-1969", in Woei Lien Chong (ed.), *China's Great Proletarian Cultural Revolution – Master Narratives and Post-Mao Counternarratives*, Lanham, Rowman & Littlefield, 2002, pp. 93-137. Sid Rittenberg was jailed for his participation as a Red Guard during the early years of the Cultural Revolution.

26 Tina Mai Chen, "Female Icons, Feminist Iconography? Socialist Rhetoric and Women's Agency in 1950s China", *Gender & History*, 15/2 (2003), pp. 268-295.

27 Shen Qian & Zhong Shi, *Shiyong meishu zi hua* [Applied calligraphy and images], Shanghai, Tongren Shushi, 1951, p. 52; Renmin Meishu Chubanshe (ed.), *Xuanchuanhua cankao ziliao huibian* [Collection of reference materials for propaganda images], Beijing, Renmin Meishu Chubanshe, 1953, pp. 31, 57, 70.

Figure 7.2. "Great Meeting"

Feng Zhen & Li Qihe, *Weidade huijian*, Beijing, Renmin Meishu Chubanshe, 1951.

Figure 7.3. "Study the Soviet Union's Advanced Economy to Build up our Nation"

Ding Hao, *Xuexi Sulian xianjin jingji jianshe womende guojia*, Shanghai, Huadong Renmin Meishu Chubanshe, 1953.

Even so, most posters devoted to the socialist support of Chinese reconstruction treat the topic with a certain degree of unease. This may have been caused by the apparent contradiction between having "stood up" while still needing foreign guidance, but it is more likely that China found it hard to bear its student status. Although the Chinese seem to be gratified that they could follow the Soviet example, there nonetheless is a feeling of distance between the represented Soviets and Chinese. The Soviet "elder brother", of course, is usually shown in the position of authority, of teacher (Figure 7.3). The Chinese are pictured as the attentive pupils, absorbing every word uttered by their teacher. As a result, most of these images do not seem to be filled with deep feelings of international proletarian solidarity, or class brotherhood. Instead, there is a pervasive feeling of Soviet condescension towards the eager but seemingly insecure Chinese pupils. The latter, in turn, do not look grateful, but rather envious. All this reflects the inherent danger of the teacher-pupil relation: that the pupil will be bullied rather than instructed, further strengthening the feeling of being slighted.[28]

Figure 7.4. "Find your Own Way to Die"

Designer unknown, publisher unknown, 1955.

… and Foes

The foes, on the other hand, consisted of all the "others", namely, the Americans and their allies, in the capitalist world, who posed a real or alleged threat to the Chinese socialist endeavour. Designers could draw upon a large and diverse body of popular sentiments to form their images. Americans not only were painted as imperialists, but also as the long-time supporters of the Guomindang (GMD) regime,[29] even before it had been vanquished by the CCP and had taken refuge on Taiwan in

[28] Gries, *China's New Nationalism*, pp. 31-35.

[29] See Barbara W. Tuchman, *Stilwell and the American Experience in China 1911-1945*, New York, Macmillan, 1971.

1949. Images of GMD leader Jiang Jieshi (Chiang Kai-shek) consorting and conniving with Americans had become regular features in all sorts of illustrated materials (Figure 7.4).[30] In the eyes and words of the CCP leadership, this American support prevented the Chinese people from realising their ultimate ambition of liberating and reuniting all the Chinese territories. Whereas the Americans initially were seen as waging a war against the Chinese through their proxies on Taiwan, the aggressive posturing of the US in the Korean War, and the direct threat emanating from this against Chinese territorial integrity, turned them into a real enemy.

To my knowledge, only a few Chinese posters have been produced that explicitly show North Koreans actively engaged in their struggle against US troops during the Korean War. In the few extant examples that are available today, the North Koreans are barely distinguishable from their comrades-in-arms from the Chinese People's Volunteers.[31] It creates the impression that the Chinese are doing the bulk of the fighting. A completely different picture emerges, of course, when we look at North Korean sources, if only because they in turn do not acknowledge any of the armed support they received from the Chinese.[32]

Chinese posters devote much attention to their involvement in the conflict, both militarily and by stressing the essentially humanitarian aspects of their actions. Interestingly, Chinese propaganda presented only a single enemy in this conflict, and that was America. No distinction was made between American soldiers and the other Western combatants assembled in the United Nations forces. Once the allegations of the American use of germ warfare in North Korea were voiced, the struggle against this became a favoured subject for posters.[33] Martial heroics were reserved for members of the Volunteers.[34] Many of the martyrs held up for emulation by the Chinese people in the early 1950s

---

30 Chang-tai Hung, "The Fuming Image: Cartoons and Public Opinion in Late Republican China, 1945 to 1949", *Comparative Studies of Society and History*, 36/1 (1994), pp. 130-131.

31 Jean-Yves Bajon, *Les années Mao – une histoire de la Chine en affiches (1949-1979)*, Paris, Les Éditions de Pacifique, 2001, p. 18; Shen & Zhong, *Shiyong meishu zi hua*, pp. 51, 53.

32 *The Path of Great Love*, Pyongyang, Foreign Languages Publishing House, 1977, pp. 60-96; *The Leader of People – Collection of Works at the National Art Exhibition in Celebration of the 70th Birthday of Great Leader President Kim Il Sung*, Pyongyang, Foreign Languages Publishing House, 1984, pp. 80-101.

33 Ruth Rogaski, "Nature, Annihilation, and Modernity: China's Korean War Germ-Warfare Experience Reconsidered", *Journal of Asian Studies*, 61/2, 2002, pp. 381-415; Renmin Meishu Chubanshe (ed.), *Xuanchuanhua*, pp. 39-46.

34 Shao Wu *et al.* (eds.), *Gongheguo qunyingpu* [Register of Heroes of the Republic], Beijing, Zhongguo Shaonian Ertong Chubanshe, 2003, pp. 14-40.

were killed on Korean soil. They included Huang Jiguang (who threw himself against the machine-gun slit of a dugout manned by American troops and died) and Yang Gensi (who, clutching a satchel of explosives to his body, threw himself into a machine-gun nest manned by Americans) (Figure 7.5).[35] As a result, posters tend to treat the spectators to images of Western casualties, although really detailed materials tend to be rare. Humanitarian actions made for good propaganda as well. Domestic support was drummed up, for example, by showing Korean grannies thanking the Chinese fighters for their heroic protection, and by innocent Korean civilians being bombed by American warplanes.

Figure 7.5. Shen Xinggong, *Yang Gensi*

Shanghai, Renmin Chubanshe, 1974.

An outright demonisation of American or European soldiers in these illustrated materials, then, is missing. Rather, recourse is sought in an adroit use of the cartoon stereotypes of Westerners that had already been popularised in the 1930s. The caricatures were obviously intended to belittle the white adversary. This time, however, they went beyond simplifications such as Uncle Sam or John Bull, and were updated with

35 Li Ming, *Huang Jiguang*, Beijing, Zhongguo Shaonian Ertong Chubanshe, 1957; Li Jingying, "Huang Jiguang", *Qingnian yingxiongde gushi* [Stories of Youthful Heroes], Beijing, Zhongguo Qingnian Chubanshe, 1965, pp. 152-179; Xu Yan, "Yang Gensi", *80 wei gongchandang rende gushi* [The Stories of 80 Communist Party Personages], Beijing, Jiefangjun Wenyi Chubanshe, 2001, pp. 249-252.

the features of contemporary American leaders and generals, including Eisenhower, Truman, McArthur, Leighton Stuart and John Foster Dulles, who are identified either as imperialists or reactionaries.[36]

Figure 7.6. Shen Tongheng, "How to Study Caricature", 1955

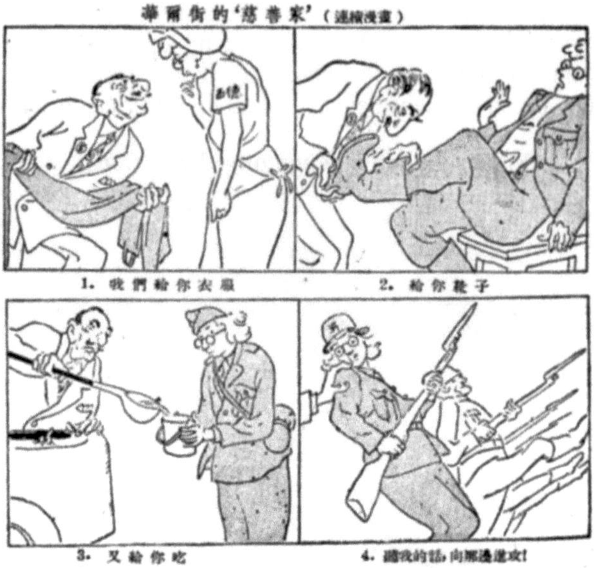

Zenyang xue manhua, Shanghai, Silian Chubanshe, 1955.

Even the economic interests of Wall Street are depicted as players in the Korean conflict, and as a result they are lampooned with the same stereotypical images used in the West of bankers in top hats chewing fat cigars.[37] There is even a suggestion of anti-Semitic inspiration in the rendering of the faces of these financiers, a technique of Soviet origin that is also applied to the American supporters of Chiang Kai-shek.[38] Such caricatures did not only appear in newspapers and publications specialised in cartoons, but also in posters (Figure 7.6).

[36] For example *Manhua* (Cartoons), 29 (April 1953); Shen & Zhong, *Shiyong meishu zihua*, pp. 19-21, 37-44. Renmin Meishu Chubanshe (ed.), *Xuanchuanhua*, pp. 15-16, 103-109.

[37] Renmin Meishu Chubanshe (ed.), *Xuanchuanhua*, pp. 101-102; Shen & Zhong, *Shiyong meishu zihua*, pp. 33-36; Shen Tongheng, *Zenyang xue manhua* [How to Study Caricature], Shanghai, Silian Chubanshe, 1955, pp. 29, 35, 43, 55, 59, 68-69.

[38] Stefan Landsberger, *Paint it Red – Fifty Years of Chinese Propaganda Posters*, Groningen, Intermed, 1998, pp. 42-43; Renmin Meishu Chubanshe (ed.), *Xuan-chuanhua*, pp. 99, 101-102.

Figure 7.7. "Collection of Reference Materials for Propaganda Images"

Source: Renmin Meishu Chubanshe (ed.), *Xuanchuanhua cankao ziliao huibian*, Beijing, Renmin Meishu Chubanshe, 1953

During the Cold War, all Western politicians played a prominent but consistently negative role in Chinese visual materials. On the British side, Winston Churchill, Ernest Bevin and others were taken to task. France's Charles De Gaulle and Germany's Konrad Adenauer were regular topics of derision, as were the other world leaders who were perceived as opposing the aims and goals of the socialist bloc with which the PRC was aligned (Figure 7.7).[39]

China *vs* the World

In the late 1950s, the Sino-Soviet split occurred and China faced a period of prolonged international isolation. Having been made the scapegoats for the disastrous economic policies of the Great Leap

[39] Shen & Zhong, *Shiyong meishu zihua*, pp. 22-25.

Forward and the famine that followed, the Soviet advisers were withdrawn. As the ideological conflict between China and the SU turned more acrimonious, the Soviets initially were not demonised, or turned into caricatures, contrary to what we might have expected on the basis of the earlier Chinese depiction of Westerners. If the Soviets appeared at all, they were taken to task in posters showing mass scenes in which the alleged Soviet revisionism was denounced with the same fierceness as American imperialism and reactionaries in every country, for which the traditional Han sense of uniqueness, superiority and contempt for Westerners easily could be marshalled. This rather favorable treatment is truly remarkable in the light of the increased tensions between the two along the shared North-Eastern border and the actual armed struggle over Damanski/Zhenbao Island (1969) that later flared up in the Amur Region.[40]

In the heat of the Cultural Revolution, when China was in the grip of a zealous attempt to revolutionise the world, caricatures of Kosygin, Brezhnev and Gromyko being crushed by the muscular arm of the dictatorship of the proletariat finally made their debut in various media (Figure 7.8).[41] Given the vehemence with which the Soviet leaders were lampooned, it must be concluded that China saw its former "elder brother" and the "Socialist Imperialism" it was said to be engaged in as a greater threat to its security than the US military presence and actions in Vietnam, along its southern border.

---

40 Lyle J. Goldstein, "Return to Zhenbao Island: Who Started Shooting and Why it Matters", *The China Quarterly*, 168, Dec. 2001, pp. 985-997.

41 *Geming da pipan baotou xuanji* [Selection of Great Revolutionary Criticism Mastheads], Hangzhou, Zhejiang Renmin Meishu Chubanshe, 1970, pp. 48-49, 52.

Figure 7.8. Nanjing Workers Fine Arts Rebel Army, "Down with Soviet Revisionism", c. 1968

Dadao Su xiu, publisher unknown, c. 1968

Figure 7.9. "Give Prominence to Politics, Temper the Skills, Attack the Aggressors", 1965

Designer unknown, General Headquarters of the People's Liberation Army, *Tuchu zhengzhi, lianhao benling, daji qinlüezhe*, 1965.

The belittling of Americans continued at the time, in line with the escalation of the conflict in Vietnam. A potential spillover of hostilities onto Chinese territory seemed both possible and imminent, all the more so as China had taken the side of North Vietnam, supporting the military and deploying a number of advisers and some troops against the American forces. The derogatory images as before included cartoon stereotypes of leaders, in particular Lyndon B. Johnson, and later Richard M. Nixon, in anguish about their military failures against the Vietnamese (and their Chinese advisers) (Figure 7.9). But the American soldier, sometimes shown as a white ape covered with black hair, surrounded by victorious members of the Vietnamese militia also, was a favourite topic.[42] The historic Nixon visit to China in 1972 started a rapprochement, not only with the US, but with other non-socialist Western countries as well. Needing the West's support, both politically and economically, to face down the threat that China felt from the Soviet Union, a less antagonistic attitude was called for. Propaganda posters, therefore, once more showed awe-struck Western Others visiting China.

Figure 7.10. "Long Live the Eternal and Unbreakable Friendship in Battle between the Peoples of China and Albania!", 1969

Revolutionary Committee of the Shanghai Red Flag Machine Factory, *Zhong Ah liang-guo renmin yonghengde, lao buke pode zhandou youyi wansui!*, Shanghai, Shanghai Renmin Meishu Chubanshe, 1969.

[42] Landsberger, *Paint it Red*, p. 49. The "Westerner as ape" paradigm remains popular to the present day. See Geremie Barmé, "To Screw Foreigners is Patriotic: China's Avant-Garde Nationalists", *The China Journal*, 34 (July 1995), p. 184.

Albanians

On the basis of the propaganda that appeared in the heat of the Cultural Revolution one is tempted to conclude that the only Western friends China had left were the Albanians. That impression, of course, is not completely correct. China had only become aware of Albania in the early and mid-1950s.[43] The Albanians, as one of the allied East-European peoples that formed the socialist bloc, had already appeared – in their distinctive national dress – in the reference materials of the early 1950s.[44] As staunch supporters of China, Albanians started to appear regularly in posters by the late 1960s (Figure 7.10). But many representatives of political splinter groups and liberation movements in Africa, Asia, Latin America and the West that subscribed to the same revolutionary values also visited China at the time, and paid homage to the Great Leader Mao Zedong while negotiating about possibilities for support beyond the ideological, and their likenesses were only seldom immortalised in propaganda art.[45]

In the late 1960s, the Chinese saw Albania as an ideal nation: it struggled fiercely against revisionism, and was successfully engaged in building socialism against all odds. Albania supported national liberation movements in the Third World and promoted the concept of armed struggle and global revolution. Like China, Albania had been a close ally of the Soviet Union and had broken away and entered into an alliance with China. The Soviet Union's ideological enemies truly were seen as China's natural friends at the time. Equally important was the fact that Albania could function as a Chinese propaganda base in Moscow's backyard. Moreover, in this relationship it was China that could play the role of "elder brother". Although no formal treaty was signed between the two nations, the alliance started to take shape from 1960 onwards.[46]

Once China began to focus less on ideological purity, and changed its stance to one where contacts with the outside world were given more weight, the Sino-Albanian alliance slowly started to disintegrate. Relations soured even further after 1976, when Albania completely misread the political developments shaking China and continued to praise Mao after his death and persevered in publicly denouncing Deng Xiaoping, who was about to make his second political come-back. In 1978, China

43 Biberaj, *Albania*, pp. 44, 64-65.

44 Renmin Meishu Chubanshe (ed.), *Xuanchuanhua*, p. 73.

45 Bates Gill & Melissa Murphy, "China's Evolving Approach to Counterterrorism", *Harvard Asia Quarterly*, 9/1&2, Winter/Spring 2005, pp. 21-23.

46 Biberaj, *Albania*, pp. 6, 48.

terminated its assistance to Albania, recalled its specialists and lost interest in images of Albanians.[47]

Historical White Personages

A special category of images is made up of certain historical white figures that were seen as having contributed to China's ideological development and thus were playing a role in the construction of the revolutionary past.[48] This category, which had already appeared when the CCP still was struggling for survival in its rural strongholds before 1949, includes the images of Karl Marx, Friedrich Engels, V.I. Lenin and Joseph Stalin, important contributors to the ideological framework that lay at the basis of CCP rule.[49] The portraits of these founding fathers of Marxist-Leninism have been ingrained in the Chinese gaze. Up to the present, they remain favourite images for use in classrooms, and they have been on display on Tiananmen Square at almost every occasion of national importance since 1949. In such displays, Marx and Engels are located on the side of the Museum of the Chinese Revolution, while Lenin and Stalin are positioned on the side of the Great Hall of the People. They all face Mao's portrait that is affixed to the Forbidden City. The portrait of Sun Yatsen, the father of the Chinese Revolution, is usually placed in the middle.

The services of these well known revolutionary luminaries were recruited for other purposes as well. The images of Marx, Engels, Lenin and Stalin often were used to lend credence to the Chinese boast that Mao was the last living theoretician of Marxist-Leninism.[50] This sentiment reached a fever pitch during the Cultural Revolution. But Marx and Lenin also made an appearance in less politically charged, more didactic posters. In 1977, a scene from the popular children's story "Lenin's Overcoat" was published, showing Lenin handing his overcoat to a sentinel. A poster produced in the early 1980s presented Karl Marx, who fathered only one daughter (Jenny), as an early advocate of the single-child family that was at the centre of the population policy adopted in 1979 (Figure 7.11). Around the same time, a poster appeared which showed Lenin distributing gifts at a Christmas tree, surrounded by children.

---

47 Biberaj, *Albania*, pp. 117-122, 135.

48 Kirk A. Denton, "Visual Memory and the Construction of a Revolutionary Past: Paintings from the Museum of the Chinese Revolution", *Modern Chinese Literature and Culture*, 12/2, 2000, p. 203.

49 Shen & Zhong, *Shiyong meishu zihua*, pp. 9-11.

50 Bajon, *Les années Mao*, p. 15.

Figure 7.11. Ha Qiongwen, "Happy Holidays", 1982

Huanlede jiari, Shanghai, Shanghai Renmin Meishu Chubanshe, 1982.

Another white foreigner prominently appearing in propaganda posters has been Norman Bethune (1890-1939), a Canadian thoracic surgeon. During the 1930s he became a convinced communist and this led him to Spain, where he joined the anti-fascist struggle. In Spain he became aware that 75 per cent of serious battle casualties would survive if operated on immediately. In early 1938, he arrived in China and proceeded to the revolutionary base area of Yan'an. Mao Zedong commissioned him to organise a mobile operating unit in the interior of North China (Figure 7.12). Although he was forced to work under extreme circumstances, sometimes operating for forty hours straight without sleep, and within minutes of the front lines, he saved the lives of many Chinese party members and soldiers. He died of septicemia, contracted when he cut himself while operating under great pressure from advancing Japanese forces.[51] Bethune was – and still is – widely admired in China. In the opening days of the Cultural Revolution (1966-1976), an early essay written by Mao Zedong entitled, "In Memory of

[51] Ted Allan and Sydney Gordon, *The Scalpel, the Sword: the Story of Doctor Norman Bethune*, New York, Monthly Review Press, 1973.

Norman Bethune",[52] was made part of the narrow selection of reading materials considered ideologically correct at the time: the courses of higher primary school students, for example, centred on them.

Figure 7.12. Zhang Fulong, "Chairman Mao Meets with Comrade Bethune", 1976

Mao zhuxi huijian Baiqiu'en tongzhi, Tianjin, Tianjin Renmin Meishu Chubanshe, 1976.

Strangely missing from this catalogue of white forebears of the CCP are the Westerners who played a role in its establishment. No attention is paid for example to the Dutchman Henk Sneevliet, the Comintern representative who from 1920 to 1924 functioned as a sort of midwife at the birth of the fledgling party,[53] nor do Sneevliet's Comintern success-

52 "In Memory of Norman Bethune", written on 21 December 1939, in *Selected Works of Mao Tse-tung*, Vol. II, Peking, Foreign Languages Press, 1965, pp. 337-338.

53 Tony Saich, *The Origins of the First United Front in China – The Role of Sneevliet (Alias Maring)*, 2 Vols, Leiden, Brill, 1991.

sors, like Borodin and others, receive any pictorial mention. The only possible explanation for this can be that the CCP was ashamed of its Western roots or of its willingness to follow directions from foreigners.

Hong Kong

Hong Kong, the former British crown colony, would have been an excellent stage on which to wage trenchant anti-Western Chinese propaganda: a piece of the Motherland, wrested away during the heyday of Western imperialism, governed by colonial, white overseers exploiting poor Chinese. The silence from Peking in this regard, however, was deafening, undoubtedly the result of the fact that for almost as long as the PRC existed, Hong Kong functioned as a window on the world, enabling the Chinese to earn foreign currency, among other things. The sole exception was the period 1966-69, when Red Guard sympathisers attempted to incite a version of the Cultural Revolution in Hong Kong. There were massive Red Guard activities which unsuccessfully tried to mobilise the Hong Kong people into joining the international historical trend of an anti-imperialist offensive against the British rulers. For these activities, a modest number of posters was published, with the added intention of drumming up popular support for the revolutionary tide sweeping within China itself. Although agitation against British suppression was the main message, much was made of the *Chineseness* of the Hong Kong people.

Figure 7.13. "Deng Xiaoping Meets with Margaret Thatcher", 1997

From Zhang Lizhen, *Zhongguo-Xianggang 1997* [China-Hong Kong 1997], 11 sheets, Shanghai, Shanghai Jiaoyu Chubanshe, 1997.

By the mid-1980s, when negotiations had started with the British about the resumption of Chinese sovereignty over the territory, Hong Kong returned as a poster topic. This time, however, it was *not* the Hong Kong people who were featured, despite one of the slogans at the time that stressed that Hong Kong would be governed by Hong Kong people. Most of the attention was devoted to the economic splendour of the region and the Chinese take-over itself, scheduled to take place in 1997. Some posters made much of the talks – negotiations would be too grand a word in this respect – between Prime Minister Margaret Thatcher and Deng Xiaoping, that sealed Hong Kong's fate (Figure 7.13). But this was clearly done to show the Chinese people how the world had changed: Westerners now played the role of eager supplicants, while China called the shots. Thus, the humiliation suffered during the so-called "Century of Humiliation" (1842-1949), in which China was forced to enter into "unequal treaties" with the Western powers and even to cede Hong Kong, could be redressed.[54]

Media Images

Since the early 1980s, television in particular has taken over the vital role of the poster to become one of the most important channels through which propaganda is waged.[55] More than ever before, this has enabled the Chinese to become acquainted *and* enculturated with images of Westerners and with models and interpretations of development and culture that often are of non-Chinese origin. Due to the fact that the CCP allowed the broadcasting of increasing numbers of television programmes produced in the West, even including the televised games of the wildly popular American National Basketball Association,[56] the Chinese, as millions of other inhabitants on the planet, have joined the "global village" – some would call it the "global mind" – of the electronic media.[57] Over the past decade, the Internet has become another powerful channel for the transmission of Western images and concepts. The rampant commercialisation of the Chinese media sphere, moreover,

54 Gries, *China's New Nationalism*, pp. 45-52.

55 Stefan Landsberger, "Learning by what Example? Educational Propaganda in Twenty-First-Century China", *Critical Asian Studies*, 33/1, 2001, pp. 567-568.

56 Andrew Morris, "'I Believe You can Fly': Basketball Culture in Postsocialist China", in Perry Link, Richard P. Madsen, & Paul G. Pickowicz (eds.), *Popular China – Unofficial Culture in a Globalizing Society*, Lanham, Rowman & Littlefield, 2002, pp. 10-12, 30-31.

57 David Morley & Kevin Robins, *Spaces of Identity – Global Media, Electronic Landscapes and Cultural Boundaries*, London, Routledge, 1995, pp. 111, 125.

offers ample space for imagined Westerners to become household familiars, both in a positive and a negative sense.[58]

Figure 7.14. Dashan publicity photograph

From: http://www.dashan.com (accessed 3 Sept. 2007).

The most famous of these mediated Westerners at the moment is undoubtedly the Canadian Mark Rowswell, known in China as "Big Mountain" (Dashan).[59] In a number of ways, the development of Rowswell's career reads like a textbook case of appropriation. Starting as an exchange student at Peking University in 1988, he soon became the Westerner the Chinese love to embrace by participating in television programmes in which Westerners are invited to show off their command of the Chinese language. These shows serve no other function than making fools of the foreigners and to state the obvious: that real Chinese can only be mastered by the Chinese. However, over the years Dashan succeeded in honing his language skills to such an extent that he has become the embodiment of the Westerner in Chinese eyes while at the same time earning a grudging respect for his commitment to China. Or, in the words of the *People's Daily*, the CCP mouthpiece: "Although Dashan is a foreigner in China, he's not an outsider".[60] All this has earned him the jealousy (and scorn) of many other Westerners residing in China, not only because his near-faultless mastering of the language makes their own attempts to acquire some degree of Chinese fluency

58 David Moser, "Media 'Schizophrenia' in China", published at www.danwei.org (accessed 7 Aug. 2006).

59 See http://www.dashan.com/en/index.htm (accessed 30 May 2006).

60 "To some in China, a Canadian is the West", at http://english.peopledaily.com.cn/200406/23/eng20040623_147314.html (accessed 30 May 2006).

look inadequate, but also because he is considered by many to be the lapdog of the government. None of this, however, has stopped Rowswell from becoming a successful media star, appearing in both serious television programmes such as *Communicate in Chinese*, *Travel in Chinese* and *To 2008*, in preparation for the Beijing Olympics, and commercials for a variety of products and causes (Figure 7.14).

In a curious way, Rowswell satisfies the Chinese need to classify and simplify. As the title of the *People's Daily* article quoted above already indicates, to many Chinese Dashan *is* the West. In other words, he has evolved from being the butt of jokes to become a visual *tifa*, a predetermined template to which a Westerner is expected to conform. Rachel DeWoskin, an American expat who unexpectedly landed the role of the "foreign babe" Jiexi in the wildly popular China Central Television soap opera *Foreign Babes in Beijing*, screened in the 1990s, recounts many such instances when she was expected to adopt behaviour in real life that her Chinese hosts thought befitted the "Western woman of loose morals" she played in the series.[61]

Rowswell's elevation to template is reflected in the work he has been offered in recent years: In 2006 he took on the lead role of the 18th century Italian Jesuit missionary and painter Giuseppe Castiglione in a 24-part television series *Palace Artist* broadcast by China Central Television, and in that same year, he performed the lead role in "Red Star Over China", a stage play based on the life of the American journalist Edgar Snow.[62] In having taken on the role of "token Westerner" in films, television programs and commercials, Dashan has filled a niche that over the decades used to be occupied by many a passing foreign exchange student from the West.

Interpreting the Other

What conclusions can we draw about the role of propaganda posters in the construction of the Chinese world view? The image of foreigners of whatever colour has been directly related to the dominant discourses of ideology and foreign relations within China itself. Although racial distinctions after 1949 no longer were seen as pointing to different levels of development but as exponents of class struggle, the distinctive hierarchic racial ordering that had been formulated and popularised in the late Qing and Republican eras continued to leave its imprint on the

---

61 Rachel DeWoskin, *Foreign Babes in Beijing – Behind the Scenes of a New China*, London, Granta, 2005.

62 See the reviews posted on Rowswell's website, http://www.dashan.com/cn/about/arts/lang/index.htm and http://www.dashan.com/cn/about/arts/redstar.htm (accessed 30 May 2006).

visualisations of the non-Chinese Other. As a result Caucasians, whether European or from elsewhere, were stuck with the label of the imperialism that had been so instrumental in breaking open China to the outside world. It did not really matter that America became the number one imperialist nation after World War II: what was important was that some peoples with white skins threatened China. More problematic was the visualisation of the Caucasian friends that China won after 1949. Designers had to do their utmost to differentiate between the dreaded Yankees and the new brothers from the Soviet Union and Eastern Europe.

Thus an uneasy and ambivalent visual repertoire was created that attempted to show how interracial relations *ought* to be on the basis of political and ideological considerations. At the same time, this repertoire was unable to camouflage the reality of the deeply ingrained feelings of Chinese racial superiority. Whenever racial conflicts break out in China, even in recent times, it is clear that such attitudes still are very much alive.[63]

On the basis of the Chinese classification of races, the white and yellow races were superior. Although the white race still had the upper hand, yellow was in its ascendancy. Proof of this was found partly in the victory of the CCP and the founding of the PRC. Unfortunately, the white race failed to give China due credit for its newfound status, and remained either antagonistic (USA) or overbearing and paternalistic (USSR). The Chinese *pique* over this is illustrated clearly by the materials that show the Americans as caricatures. By ridiculing them, it might be possible to gloss over the fact that Americans could be considered as superiors. The posters devoted to the Soviet "elder brothers" in the 1950s, on the other hand, illustrate how difficult it was for Chinese to swallow their racial pride. These images were intended to make the Chinese population accept the political reality that forced it to learn from representatives of a people that it basically considered inferior in cultural accomplishments. But at the same time, the images failed to mask the resentment – undoubtedly deeply felt – that China had fallen to this level. It is small wonder, then, that once the Sino-Soviet split occurred in 1960, Soviets too were caricatured. The relations with the Albanians were much more to China's liking because it could occupy the senior position in them. Although the propaganda posters presented the Albanians as equals in the struggle against Soviet revisionism, the Chinese probably all knew better.

63 Barry Sautman, "Anti-Black Racism in Post-Mao China", *The China Quarterly*, 138, June 1994, pp. 429-434.

If these propaganda posters devoted to foreigners were not intended to teach the Chinese people about their place in the world or to help them identify with other downtrodden peoples, why then were they produced at all? The answer must lie in the fact that they were published mainly for domestic purposes, and for Chinese audiences. The materials that showed various ethnically diverse, hard-to-distinguish armed freedom fighters in mass scenes waving copies of *The Thoughts of Chairman Mao* to guide them in their struggles, served only a single function. They were meant to strengthen the impression that China, as the champion of the downtrodden, was leading other races in the global struggle against white imperialism. Moreover, it gave another boost to the claims that *The Thoughts of Chairman Mao* had universal applicability. These considerations in turn were supposed to give the Chinese an added feeling of superiority.

With the decline of the Chinese propaganda poster after the late 1970s, the Other made his debut on television, becoming a fixture there and in other electronic media. Politics are dead, and consumerism is very much alive. The continuing irrelevance of politics has reduced the effect of any propaganda practically to nil, and has given way to other media that present us with a more fluid, contested relationship between conceptions of modernity and evaluations of foreign practices. Under these conditions, and with multitudes of non-Chinese, including Westerners, having entered the country, mediated interpretation of the Other is no longer necessary.

Moreover, the Chinese people themselves have become more obsessed with "becoming white" – and thereby superior, in particular from the late 1990s onwards (Figure 7.15). The pale look – traditionally seen as defining one's membership of the aristocracy, but later indicating that one was not forced to work outside the house, like a peasant – has become a standard that many, women in particular, try to achieve. And for those who for whatever reason fail to realise this goal, many Chinese and Western companies offer their skin-care products through television commercials and print advertising. White skin is seen as a signifier for status, as one of the markers of belonging to the newly emerging "middle class" of urbanites, rather than to the ever-growing body of migrants living and working in the cities without a permanent residence permit. The commercial propaganda that inundates Chinese viewers on a daily basis shows how to attain it.[64]

[64] Perry Johansson, "White Skin, Large Breasts: Chinese Beauty Product Advertising as Cultural Discourse", *China Information*, 13/2&3, Autumn/Winter 1998, p. 61; Wang Xiaohua & Sun Qing, *Bainian shenghuo bianqian* [A Hundred Years of Changes in Life], Nanjing, Jiangsu Meishu Chubanshe, 2000, pp. 185-190.

Figure 7.15. "Be Filial to your Parents", 1996

Huang Hong'en, *Xiaoxin xiangei fumu*, Kexue Chubanshe, 1996.

CHAPTER 8

Arab Windows on Europe

Robbert A.F.L. WOLTERING

Even if they are aware of the fact that the great majority of Muslims are not Arabs, Europeans cannot escape the fact that the part of the Muslim world most relevant to their semi-continent is the Arab world: that string of countries stretching from Morocco to Iraq. The Arab world, the Muslim region nearest to Europe, supplies it with oil, the lifeblood of its economy. As regards a major issue in contemporary European politics, immigration, there are but few immigrants making their way to European shores without at some point crossing Arab lands. Similarly, European concerns with Islamic radicalism and terrorism are focussed on the Arab world, or on Arabs living in Europe. As of the 1970s the Arab world has witnessed a rise in the political relevance of religion, and Islamism has replaced Arab nationalism as the primary concern of European monitors of the Arab world. All this makes it understandable that there is a certain cross-over between "Arab" and "Muslim" in the European imagination. This European confusion between Arab and Muslim is paralleled in the Arab world: Arabs know the difference between Europe and the West, but these two terms are often equally interchangeable. And perhaps the parallel runs deeper: after all, where Europe spawned the Western world, Islam sprung from Arabia. In that limited sense, we may all be excused for our *pars pro toto*s and metonyms.

And yet, for a proper understanding, we are forced to make distinctions. In this chapter I will focus on the various views of Europe as presented in intellectual debates in the Arab world, particularly in Egypt. Only incidentally I will venture beyond the Arab world into the wider Islamic world. I will begin with a review of existing European preconceptions of the Arab view of Europe, and see to what extent these preconceptions square with the historical relations between Europe and the Arab world. Consequently, an analysis of the image of Europe in a number of recent Arab publications will be provided. We will see that the themes, stereotypes and function of contemporary Arab constructions of Europe have fascinating parallels historically and geographically. Finally, we will also see that Europe is very often discussed in its relation *vis-à-vis* the United States, and in comparison with the ideal of Arab unity.

Preconceptions

In early 2006, Danish flags were being burned in various national and regional capitals in the Muslim world, accompanied by slogans extolling the greatness of God and calling for the death of Denmark. However, the images of ferocious Muslims acting aggressively towards a Western enemy were a common sight even before the Danish cartoon drama began. They fit a common European understanding of what *they* think of *us*. Generally, Arabs and Muslims are thought to stand opposed to the West, often because their culture and religion are taken to be irreconcilable with Western values. Throughout the European Union there are spokesmen and women for this view.

After September 11 the veteran Italian journalist Oriana Fallaci committed herself to sounding the alarm over what she perceived as a threat to Western civilisation comparable to Nazism. "You don't understand, you don't want to understand, that a Reverse Crusade is underway. A War of religion they call Holy War […] You don't understand […] that for those Reverse Crusaders the West is a world to conquer and subjugate to Islam".[1] By "them", Fallaci means not a specific group of radicals, but rather Muslims in general.[2] Similarly, the Somali-born Ayaan Hirsi Ali, former Dutch conservative liberal party Member of Parliament and regarded by many as champion of Muslim women's rights, has devoted her life to waging a "liberal jihad" against "genuine Islam. Her concern is not with radicals, but with Muslims in general, for in her view, even those Muslims who may appear to have adjusted to European norms may well be fundamentalists waiting to emerge.[3]

Among European intellectuals and politicians we can find others who espouse the view that Europe is under threat by hordes of fanatic Muslims who are hateful of European culture (one may think of the sizeable extreme right parties in Denmark, Belgium, France and

1 Oriana Fallaci, *The Rage and the Pride*, New York, Rizolli, 2002, p. 27.

2 "Extremist fringes?!? Fanatic minorities?!? They are millions and millions, the extremists. They are millions and millions, the fanatics". *Ibid.*, p. 29.

3 "Westerners see in a Muslim girl in a skimpy shirt or in a rapper [of Moroccan descent], Muslims who have left orthodoxy behind them. But the Muslim girl and rapper actually have orthodoxy ahead of them. […] At a later moment in their lives, they will eventually start abiding by the rules of Islam, and [through] the rhetoric of the caliphate, they can easily become practising fundamentalists". Ayaan Hirsi Ali, "Strijd tegen absolute Allah is geboden", *NRC Handelsblad*, 8 Nov. 2005.

Austria). In the academy these views are harder to come by. Perhaps the Spanish professor of Arabic studies Serafin Fanjul is one of the few: "All [those Muslims], at some point, accuse us of islamophobia, of intolerance and racism because we don't applaud our annihilation. In time, they achieve several of the crucial objectives of Muslim society: the occupation of the public space, with the immediate exclusion of the [non-Muslim] rest".[4] For a more comprehensively worked out theory of how and why the Muslim world supposedly detests Western civilisation, we need to turn to the British scholar Bernard Lewis.

The gist of Lewis' understanding of the Arab (-Islamic) world's view of Europe and the West centres on a model of primordial conflict between Islam and the Judaeo-Christian world. Since its inception, Islam has been an enemy of non-Islam and specifically of Christianity as its main rival in monotheism. For many centuries Islam held the upper hand, enjoying a superiority in culture and conquest. But then came the European Renaissance, with which began the rise of what would gradually come to be known as "the West". Throughout the centuries that it took European societies to build modern civilisation, the world of Islam remained fatally disinterested. Lewis notes in numerous publications that very few Muslims visited the infidel lands to the West, and that those who did visit failed to see the importance of what was happening on the continent. The Muslim world view entailed that Muslims are superior to non-Muslims, and therefore the thought that non-Muslim societies could have something that Muslims should learn from, was unthinkable. And so the world of Islam declined, while its adversary grew stronger. Only when this growing discrepancy manifested itself on the battlefield did Islam (actually: the Ottoman empire) allow for changes. However, these changes were confined to practical matters of the highest immediate urgency: military material and techniques. It was not until well into the nineteenth century that the Muslim world started to show a serious – though still grudging – interest in Europe as a civilisation from which one could and should learn. According to Lewis, the uncomfortable reality that Muslims have been overtaken by the infidel West still plagues the Muslim world.[5]

Lewis' view of Islam as having been at the throat of Judeo-Christian Europe for fifteen centuries is rather one-sided and appears to be the result of selective reading. Throughout history there have been alliances

4 Serafin Fanjul, "Islamofobia", *ABC*, 24 Feb. 2006.

5 Lewis expounded this view in various publications and articles, most recently in *From Babel to Dragomans: Interpreting the Middle East*, Oxford, Oxford UP, 2004, mainly chapters 13, 21 & 33. See also his *What Went Wrong? Western Impact and Middle Eastern Response*, Oxford, Oxford UP, 2002.

between European and Muslim powers (mainly with the Ottoman empire and semi-independent Morocco).[6] Similarly, the Andalusian period (711-1453) was as much a period of mutual gain between Christian Europe and the world of Islam as of military strife. What is more, Lewis' presentation omits the fact that Christian European states have more often fought one another than any Islamic foe. Most importantly for the purposes of this chapter, the suggestion that the Muslim reaction to the emergence of modern Europe was one of sheer indignation, trauma and outright rejection, is not corroborated by the most important historical texts, as will be clarified below.

Reform in the Face of Modernity

It is widely recognised that the Arab world is in dire need of reform.[7] What is easily overlooked is that the Arab world is no stranger to reform. The Arab world did not remain stagnant and "traditional" in the face of European dominance, but rather has engaged in modernisation ever since the early nineteenth century. Just a few years after French troops crushed the Mamluke soldiers in the battle of the Pyramids, Muhammad Ali of Egypt set course for the "modernisation" of the Egyptian army, education and administration. Modernisation, in this case, meant reform after European models. One religious scholar whose writings pointed in this direction was Abd al-Rahman al-Gabarti (1754-1829).[8] Al-Gabarti opens his chronicle of the French occupation of Egypt with a poetic description of shock at the infidel victory over Muslim Egypt, but what follows is not the story of an embittered, traumatised man. Al-Gabarti remains not only factual in his account, he displays genuine interest in the French ways. What is more, Al-Gabarti even praises the French for their valour in battle, for their science and for their justice system. Clearly, this first Muslim Egyptian reaction to a crushing defeat and subsequent occupation by a European army cannot be described as particularly indignant, traumatic, or dismissive.

Part of Governor Muhammad Ali's ambitions to modernise Egypt consisted of sending students off to European capitals to study in the best universities. Rifa'a Rafi'i al-Tahtawi (1801-1873) was one of those students. Al-Tahtawi is a crucial figure in the intellectual phase which

6 Lewis does mention these alliances, but states that "it would be an exaggeration to call [it] an alliance", the flimsy reason for this being that the alliance was not on equal terms (Lewis, *What Went Wrong*, p. 22).

7 See the calls for reform as expressed in the *Arab Human Development Reports* published by the UNDP in 2002, 2003, 2004 & 2005.

8 S. Moreh, *Al-Jabarti's Chronicle of the First Seven Months of the French Occupation of Egypt*, Leiden, Brill, 1975.

has been labelled the "first generation" of Arab thinkers engaging with modern Europe, i.e. the generation between 1830 and 1870.[9] In this period Arab *exposés* of Europe were intended to inform the Arab world of the new era of development and progress. Al-Tahtawi, and others of his generation such as the Tunisian Khayr al-Din and the Lebanese Maronite Butrus al-Bustani, gained knowledge of the scientific and societal achievements of Europe, and wrote to inform a public that was largely ignorant of this.

Al-Tahtawi's account of his stay in Paris was an important book, for it provided the Egyptian reader with a first-hand report of what Europe and France in particular had to offer. Also, it provided a posture to adopt in the face of European supremacy in so many fields. *Takhlis al-ibriz fi talkhis baris*,[10] which could be loosely translated as "Extracting the finest lessons from a summary account of Paris" was intended to educate a wide and diverse Egyptian readership. *Takhlis* begins with the descriptions of leaving Cairo for Alexandria. Although Al-Tahtawi had never been to Europe, Alexandria strikes him as European. Not only are there many foreigners and do the people speak a kind of Italian (then the Mediterranean *lingua franca*), but also the buildings are to him quite unlike anything anywhere else in Egypt. Why would Al-Tahtawi write this? We know from a subsequent passage that he considers the Europeanness of Alexandria a good thing, which suggests that this is a way of telling the readers that Egypt is capable of becoming like Europe, and that it probably should.

Al-Tahtawi provides two divisions of the world, one based on civilisation and another on Islam. The first division is an explanation of the differences between the societies that are barbarian, the societies that are highly civilised, and those societies in between these. Here, Al-Tahtawi expresses the idea of progress, and surprisingly, he suggests progress is made through man-made knowledge. God does not come into play in his division between barbarity and civilisation. For instance, the first step from barbarity to a higher form of human organisation is made when a caveman notices how a falling rock falls on another rock, creates a spark and causes fire. If the man then understands that he can use this knowledge to his advantage and make his own fire, advancement has been made in his society. Similarly, higher stages are reached, apparently without God ever doing anything. The second division of the world is based on Islam. The Asian continent ranks highest, because here Islam

9 Albert Hourani, *Arabic Thought in the Liberal Age, 1798-1939*, Cambridge, Cambridge UP, 1983, p. vi.

10 Available in translation as: Daniel L. Newman, *An Imam in Paris: Al-Tahtawi's visit to France (1826-31)*, London, Saqi Books, 2004. Page references are to Newman.

has its roots and its largest group of followers. Second is Africa, because all of North Africa is Muslim, as are some regions further south. Third is Europe, because its Eastern regions form part of the Ottoman Empire, even though most of the people there are not themselves Muslim. America ranks last, for there, there are no Muslims at all. If the objective was to list the four continents according to the number of its Muslims, it would have been a rather pointless exercise. But the point of this listing is made when Al-Tahtawi illustrates what this list does *not* mean: "The above has been established from the point of view of Islam. [...] One cannot say that most if this relies on superiority, since by itself it [i.e. "Islam"] does not lead to excellence and virtue".[11] The listing was thus intended to tell the readers that Islam and superiority, excellence and virtue do not necessarily go together.

In general, Al-Tahtawi is very pleased with what he sees in Paris. First and foremost, he is taken in with the scientific advances, and states unequivocally that "The best of science is in the land of the Franks".[12] Also, people are very clean, transport and the mailing system are incredibly effective, and people strive to improve their work constantly. Al-Tahtawi praises the benefits of newspapers to educate the people, the intelligence of the people (even of the common people), and after a visit to a royal palace he concludes that "The basic principle with the French is that everything is done for the sake of beauty and elegance, rather than for [excessive] ornamentation, the outward show of wealth or vainglory".[13]

As for the negative qualifications Al-Tahtawi applies, they are few and far between. The search for profit may have beneficial effects, but it also leads to greed. In general, the French are not generous, but Al-Tahtawi adds a few things that could explain this: "Admittedly, generosity is quite rare in civilized countries... Also, they believe that giving something to someone who is capable of working induces him not to concern himself with earning a living".[14] Elsewhere, Al-Tahtawi explains the French lack of generosity by saying that "generosity is peculiar to Arabs".[15] The way the French treat "their women" is also not to Al-Tahtawi's liking, as the men are like slaves to their women, and he

11 *Ibid.*, p. 125.

12 *Ibid.*, p. 99. Unlike Al-Gabarti, who writes of *al-Faransis* (the French), Al-Tahtawi uses the old term "Franks". Possibly he preferred to use a term that his audience was familiar with: after all, he was already presenting the audience with so much that was new to them.

13 *Ibid.*, p. 218.

14 *Ibid.*, pp. 238-239.

15 *Ibid.*, p. 175.

quotes a French source saying that "The French treat [women] like spoilt children". Yet the lack of jealousy means not lack of honour, for indeed "when their women misbehave, they are the most malicious of men against themselves and against those who have betrayed them with their women".[16] Of course there was also the general flaw in the French in that they were not Muslims.

Perhaps the only serious problem Al-Tahtawi had with his Parisian experience was that some of the finest minds in Paris believed that there was no God at all, and considered religion a dangerous figment of the imagination. These are "extreme rationalists" who believe that scientists are more useful than prophets.[17] As much as Al-Tahtawi was enthralled with the rationalist approach to science, he could not allow for atheism.

In sum, Tahtawi's appreciation of Europe cannot be squared with the theory that Muslims found it hard to learn from non-Muslims. Neither does Tahtawi limit his recommendations of Western culture to material things. Rather, he is quite open-minded and enthusiastic about both material and immaterial elements of European life. In the *Takhlis* we mainly find images of a Good Europe, from which Egypt can and should learn. In the fields of education, science, public administration and the judiciary, Egypt should look to Europe.

Reform in the Face of Alienation and the Imperialist Other

Towards the end of the nineteenth century, increasingly forceful imperialism and Egyptian and Arab nationalism complicated the positive Arab views of Europe. In 1882, Egypt had to cede its financial sovereignty to England and France. England was to play an increasingly important role in the decades ahead. Soon after, mixed opinions of Europe and Europeanisation emerged. One of the most important intellectuals of the late nineteenth century was the Egyptian Islamic scholar Muhammad Abduh (1849-1905). Abduh agreed with the Egyptian nationalists that the English should leave Egypt, but more importantly, he promoted internal reform. He simultaneously aimed at making Islam compatible with the changed and changing world, and making the process of change conditional to precepts of "true Islam". This meant that Islam had to be released from the shackles of tradition and stagnation (*taqlid* and *gumud*) and that society and its modernisation had to be made subject to a set of as yet to be formed Islamic principles.[18] Con-

16 *Ibid.*, p. 362.

17 *Ibid.*, p. 180.

18 Hourani, *Arabic Thought*, p. 135.

trary to Al-Tahtawi, Abduh was very much aware of some dangers of the increasing trend of Europeanisation.

Justice was one of Abduh's main concerns, just as it was for Al-Gabarti and Al-Tahtawi before him. Abduh stressed that there could be no salvation in importing European laws to Egypt, because specific laws are based on specific historic trajectories and principles commonly understood by those who are governed under these specific laws. In other words, societies must engender their own laws if they are to be just societies. Yet Abduh did not shy away from being inspired by European sources, and went to Europe as often as he could "to renew himself".[19]

In *Risalah al-tawhid* ("The message of unity"), one of Abduh's main publications, we find an interesting historical reference to Europe. Abduh mentions the Crusader period, and asks, "why did they come, and with what did they leave?". He tells of how the "Westerners" were initially filled with hatred of Islam, indoctrinated as they had been by religious leaders in Europe. Yet as they dwelled in the Levant, surrounded by Muslims, they found that reality did not correspond to their prejudices. The Westerners found Islam to go hand in hand with freedom of religion, science, justice and art, and that freedom of thought and the quest for knowledge were means of the religion, not adverse to it. They learned from the Muslims, and not long afterwards, their ideas changed, and the desire to learn grew among the Westerners. The desire to break the chains of tradition emerged. Not long after, Protestantism emerged, as did various smaller sects. These schools of religious thought, Abduh stresses, were either very similar to Islam or even the same as Islam except for the acceptance of the role of Mohammed.[20] In this small discussion of the Crusader episode,[21] there is much of relevance to our search for images of Europeans. Abduh suggests that Europeans, before they came into contact with Islam, were rather uncivilised, hate-mongering fanatics. Only when they met with the refined culture of Islam did their minds open to freedom, science and arts. Here, Abduh essentially presents European civilisation as one that was spawned by Islam: a strategy aimed at reconciling oneself with the differences between a weak self and a dominant Other.

---

19 *Ibid.*

20 Muhammad Abduh, *Risalah al-tawhid*, Al-Qahira, 1953, p. 193-194.

21 Abduh does not actually use the term "Crusaders"; rather he refers to *al-gharb* and *al-gharbiyyun*. This use is clarified by Muhammad Rashid Rida, editor of the 1953 edition, in a note: "This refers to the Crusader wars [aimed] at driving out Islam […]. It is necessary for every Muslim to know of its details and of how the Europeans profited from [experiencing] the virtues of Islam which led them to reform the affairs of their religion and their state. Most Muslims are ignorant of this" (*Ibid.*, p. 192, n.2.).

In the early decades of the twentieth century, the modernist stream of thought split up into separate intellectual currents, each developing its own position *vis-à-vis* Europe. A major cause of worsening relations was the First World War and the subsequent developments (the Sykes-Picot Agreement). For Arab intellectuals it became difficult to remain unequivocally enthusiastic about Europe. Muhammad Rashid Rida, a key figure in early Islamism, described this disenchantment in no uncertain terms: "Europe has destroyed all the good reputation it had in the East [...] Nobody believes the word of the Europeans anymore, nor does anybody trust them or even perceive them to be qualified to exercise justice and virtue".[22]

After the First World War, the two most prominent ideologies to emerge from the modernist trend of the nineteenth century were Islamism and nationalism. Nationalism in Egypt initially meant Egyptian liberal nationalism, which espoused the view that Egyptians were closer to Europeans than to Arabs. This Egyptian nationalism later gave way to socialist pan-Arabism as championed by Gamal Abdel Nasser. The Islamist current has witnessed major internal ideological shifts as well, varying from following a nationalist course to aiming at re-establishing the caliphate, from being mainly anti-Communist to being mainly anti-Western, and from qualifying itself as either "fundamentalist" or *mu'tadil* ("balanced", often translated as "moderate").

Three Stereotypes

In the preceding overview of nineteenth and early twentieth-century Arab reactions to Europe, we can discern certain distinct Arab identifications of the European Other, all of which contain (implicit) identifications of the Arab self. These identifications or images of Europe may best be described as "stereotypes".

The major stereotype to emerge in the nineteenth century was that of Europe as an ideal for Arabs to aspire to. When we read that Tahtawi views Europeans as orderly and advanced in science, this implies that he understands the Arabs to display a relatively chaotic and scientifically backward society. When Tahtawi states that in France even the poor can read, he provides an implicit contrast with the illiterate masses of Egypt.

A second stereotype of Europe is that of Europe as immoral. The image of immoral Europe we find in Rashid Rida's qualification of Europe as treacherous and deceitful. Europe may be militarily strong, and able to produce amazing products, but all this is merely the veneer of a

22 Quoted in: Emad Eldin Shahin, *Through Muslim Eyes: M. Rashid Rida and the West*, Herndon Va, International Institute of Islamic Thought, 1993, p. 85.

corrupt civilisation. In this stereotype, the concept of authenticity becomes central. Whereas Tahtawi thought mainly of progress, Rashid Rida and others like him were concerned with the integrity of Arab culture and Muslim identity. In the face of a rapidly Europeanising society, they wanted to create a space which would be immune to the encroachment of Europe-inspired changes, a space which could counter the European assumption that Oriental peoples are the inferiors of Europeans. The ensuing image of the European immoral Other was specifically created to contrast with the superior "authentic" Arab, Egyptian, Muslim self. This image is by no means peculiar to the Arab world. Much of the Arab critique of the presumed immoral West can be found in the works of European and Russian authors.[23]

A third stereotype of Europe could be called "appropriated Europe". The European imagination promoted the idea of a progressing West and a stagnant Orient as two distinctly separate entities, that could never meet, and hence would always be unequal. As this view spread in the colonies, protectorates and beyond, intellectuals in countries as diverse as Egypt, Turkey, Russia and Japan developed strategies to find a way around this rather unpleasant axiom. In Turkey, Ziya Gökalp (1875-1924) rewrote both European and Turkish history in order to show that Europe and Turkey had a common (Turkic) ancestry. In Japan, reformers argued that Japanese society was not really "Oriental", but more similar to societies of the West. In nineteenth-century Russia, intellectuals were divided between "slavophiles" who propagated the view that "the Russian" has a distinctly Russian essence which made him superior to the European, and "Westernisers" who claimed that Russia was an organic part of Europe.[24] In Egypt, Khedive Ismail challenged the notion that "East is East" when he exclaimed in 1978, "Mon pays n'est plus en Afrique, nous faisons partie de l'Europe

---

23 For Russia, one should think of the debates between "slavophiles" and "Westernisers" (as described in Willem G. Weststeijn, *Russische literatuur*, Amsterdam, Meulenhoff, 2004, p. 35-40); we could also point at Spain, another – in this case former – "margin" of Europe: Juan José Lopez Ibor analysed the Spanish inclination to couple Spain's comparative material weakness with a presumed superior Spanish morality (*El Español y su complejo de inferioridad*, Madrid, Rialp, 1951). As for the European centre, we can point to Oswald Spengler's *Untergang des Abendlandes*, Munich, Beck, 1918-1922, in which the author predicted the demise of Europe due to a lack of morality. The phenomenon whereby intellectuals from the Muslim world oppose the West with arguments that have a Western counterpart (or even a Western progeny), has been noted earlier by Daniel Pipes ("The Western Mind of Radical Islam", *First Things*, 58 (1995) pp. 18-23), and more recently by A. Margalit & I. Buruma, *Occidentalism – the West in the Eyes of its Enemies*, New York, Penguin Press, 2004.

24 See previous note.

maintenant".[25] The idea that Egypt was more European than Oriental, African, or even Arab, continued for decades and found its apex in Taha Hussein's treatise *The Future of Culture in Egypt*, published in 1938, in which he declared that the Egyptian mind is European, rather than Oriental. Similarly, Muhammad Abduh stressed that modern Europe only emerged after it had been given the ingredient for its progress by the Muslim societies it encountered during the Crusader period. All these intellectual endeavours sought to establish a single argument: if it is true that the European Other is strong, this is because it has much in common with the self. In that sense, Europeanisation is in many ways simply a process of retrieving one's self. Or, to the extent that Europe is commendable, it is not the "Other", but rather it is the authentic, more advanced self.

Contemporary Views of Europe

It is extremely difficult to categorise the intellectual landscape in the Arab world today. The long-established lack of political freedoms has stifled the proliferation of credible political opposition parties. The lack of press freedoms smothers the display of intellectual debates which otherwise could be monitored freely. A conventional division employed both within and outside the Arab world is between "Islamist" and "secular" discourses.[26] Others have opted for a more comprehensive categorisation comprising Islamic, Marxist/leftwing, nationalist and liberal intellectual trends.[27] In this chapter I will present and discuss three recent publications by three different authors, all belonging to different trends. While these publications each fall within a certain discourse ("left-wing nationalist", "liberal" and "Islamic" respectively), they cannot of course serve as comprehensive illustrations of those discourses. The following *exposé* merely intends to highlight a number of common strategies in Arab appreciations and representations of Europe and the West.

Galal Amin

In 2002 the first Arab Human Development Report was published. This UNDP-sponsored report, drafted by a team of prominent Arab scholars and intellectuals, described with great candour the major prob-

25 *Moniteur Egyptien*, 24 Aug.1878.

26 E.g. Nadje al-Ali, *Secularism, Gender and the State in the Middle East: the Egyptian's Women Movement*, Cambridge, Cambridge UP, 2000.

27 Ibrahim M. Abu-Rabi`, *Contemporary Arab Thought: Studies in Post-1967 Arab Intellectual Thought*, London, Pluto Press, 2004.

lems in the contemporary Arab world. While the report was praised by many for its clear analysis of the three deficits in the Arab world (freedom, knowledge and women's emancipation), some vehemently opposed it for serving Western interests. Rather than elaborating on what was wrong with the Arab world, the authors should have focussed on the oppressive role of external actors. By neglecting to do so, the report was only of service to "the West".[28] One of the more prominent Arab intellectuals to attack the report was the Egyptian economist Galal Amin of the American University in Cairo. For Amin, the report almost constituted treason, since it played perfectly into the hands of the West. Amin was not alone in this assessment. Various nationalist and leftist newspapers and intellectuals criticised the report for the same reason. In one of Amin's recent publications we can see how strongly he feels about the importance of the West, and how he identifies the West as an imperialist entity much in the sense that some would think is outdated in our time. In *The Era of Arab and Muslim-bashing. We and the World after September 11*,[29] Amin argues that the Arab and Muslim world is suffering from a global smear campaign. For instance, he argues that the "events" of 11 September 2001 are thought by many to have been organised by Muslims, while in reality this was not the case. Amin does not elaborate as to who was behind the attacks if not Al-Qaeda, though he hints at the likelihood of an "inside job". His main aim, however, lies in alerting the Arabs and Muslims to the smear campaign against them and "reminding them of the justness of their case".[30] In *The Era of Muslim-bashing* Amin rejects everything that does not fit the image of a noble, well-to-do Arab world. To Amin, the images of plunder in Baghdad (right after the American invasion) must have been part of the greater ploy to defame Arabs, for the footage of looting Iraqis fitted the profile of the "barbaric Arab" too well. The image of the West as presented by Amin remains rather vague, which is perhaps necessary if it is to be blamed for the wide range of wrongs in the world that Amin mentions. Perhaps unsurprisingly, since Amin is an economist, the West is defined in economic terms: capitalism and imperialism are the hallmarks of the

[28] In a way, Amin was right in his assessment that the report would be of service to "the West". The American "Greater Middle East Initiative", which proposed a major remapping of the Middle Eastern political landscape, argued that the Arab Human Development Report showed the need for the initiative. This was much against the intentions of the authors of the report, and the lead author commented that "[the Americans] are using the report in the way a drunk leans against a lamppost; not for being enlightened, but in order not to fall down". See R. Woltering, "Ik dacht dat de UNDP het nooit zou publiceren", *ZemZem*, ½, 2005, p. 9.

[29] Galal Amin, *'Asr al-tashhir bi-l-'arab wa-l-muslimin nahnu wa al-'alam ba'd 11 sabtambar 2001*, Al Qahira, Dar al-Shuruq, 2004.

[30] *Ibid.*, p. 6.

West. The end of the Cold War has left the world with a new situation, Amin argues, in which capitalism has re-emerged in its original crude form. While during the Cold War capitalism was diluted, our present era is characterised by the kind of capitalism that was prevalent when Karl Marx lived.[31] Amin is far from alone in levelling his critique against the West in economic terms: with the demise of the Soviet Union and the Communist alternative to liberal capitalism, it might seem that capitalism is no longer challenged and that anti-Westernism is a mainly culturally informed antagonism. Amin and other left-wing Egyptians suggest that this is not the case.

Muhammad Imara

Over the past years increasing attention has been paid to so-called "moderate Islamists" or "Islamic reformers" sometimes referred to as Islamists of "the middle way", as distinguished from "fundamentalists".[32] It is suggested that this trend has emerged from within Islamic fundamentalism, and contrary to their radical progenitor, the Islamists of the middle way are considered to be willing and able to compete in a liberal democratic setting. Undeniably, the Muslim world cannot be divided in a black and white manner, with undemocratic fundamentalists one the one hand and democratic secularists on the other. Yet one should be careful in imagining that there is a clear Islamic "third way", similar to e.g. the Christian-Democratic parties one finds in Europe. Rightfully or not, Muhammad Imara is one of those Islamic thinkers commonly categorised as "moderate Islamist" (in Arabic the word used is *mu'tadil*, meaning "balanced" rather than "moderate"). Through his various weekly columns and appearances on television he is an important contributor to Arab public opinion. In a recent publication Imara focussed on the relations between Islam and the West.[33] He takes it as a matter of fact that the relations between Islam and the West have always been at best problematic, and at worst, bloody. Much as some Western authors tend to think of Islam as a civilisation that, since its inception, has been an opponent of the West (or Europe, Christianity or the "Judaeo-Christian civilisation"), Imara portrays the West as a civilisation that has never really changed its position regarding the Orient (or

---

31 *Ibid.*, p. 10.

32 Two recent publications in which a particularly jubilant tone is employed are Raymond Baker, *Islam without Fear. Egypt and the New Islamists*, Cambridge Mass., Harvard UP, 2003; and Geneive Abdo, *No God but God. Egypt and the Triumph of Islam*, Oxford, Oxford UP, 2000.

33 Muhammad Imara, *Al-islam wa al-gharb. Ayna al-khat'a?.. wa ayna al-sawab?*, Al-Qahira, Maktaba al-shuruq al-dawliyya, 2004.

Islam, or the Arabs). The West, according to Imara, has always been imperialist: "The history of Western imperialism in the East preceded the coming of Islam. The Roman and Byzantine West had created in Egyptian and Eastern Christianity and in the [...] civilisation of the Orient, the false idea that condoned oppression [...] at the hands of the Romans and the Byzantines. [This oppression] of the Orient and the Oriental peoples [lasted] for ten centuries, from Alexander the Great [...] to the seventh-century [Byzantine emperor] Heraclius".[34] All this only ended, Imara then explains, when Islam came onto the scene and liberated the Orient from Western oppression. From that moment on, the West has been at odds with Islam. This formulation of history serves two purposes. Firstly, it provides the reader with an image of the West as imperialist *per se*. Secondly, and partly as a consequence of the first lesson, there is nothing Islam can do to alter the bad image that the West has of Islam. The West already exercised its oppression over the Orient before Islam emerged, and this is proof that the West would be aggressive towards the Orient regardless of Islam. The point is also made elsewhere in the book, when Imara stresses that the West campaigned against Islam long before September 11, and that clearly September 11 cannot be seen as the justification for Western attacks upon Islam. Imara's portrayal of the West is different from that of Amin in the sense that Imara focusses more on the idea of a religious conflict, in which the West stands opposed to Islam as a revolutionary religion and as an alternative to Western secular political systems. Though Imara clearly has grievances against the West of an economic nature, he stresses the cultural and civilisational differences between the Western and the Islamic worlds. In that sense his discourse fits Huntington's theory of a "clash of civilisations", as well as Bernard Lewis' representation of East-West relations as characterised by perennial conflict.[35]

Rida Hilal

Whereas the concept of "the West" includes Europe and North America, ever since the end of World War II it is the United States which has commonly been viewed as its primary component. The liberal intellectual Rida Hilal goes further and tends not to speak of the West or Westernisation, but rather of America and Americanisation. In his recent

---

34 *Ibid.*, pp. 65-66.

35 Huntington actually derived the phrase "clash of civilisations" from Bernard Lewis' article "The Roots of Muslim Rage", first published in 1991, collected in his *From Babel to Dragomans: Interpreting the Middle East*, Oxford, Oxford UP, 2004.

book, *Americanisation and Islamisation*,[36] Hilal calls for a marriage between Islam and democracy, to take the place of the current development in which rigid Salafi Islamisation is coupled with American consumerism. Interestingly, Hilal contrasts the current process of Americanisation with the pre-World War II process of Europeanisation. Whereas today the Arab world is taking only the froth of America, leaving aside its true riches, which are its values, Hilal argues that the previous age of Europeanisation was much more constructive: "In the period of Europeanisation [...], Egyptians took from the Europeans the rule of law, constitutionalism, parliament, journalism, theatre and cinema".[37] The result was that Egypt could vie with the best of the world's nations, and Cairo could stand up to Paris or London. This so-called Renaissance-period, Hilal argues, was eventually frustrated by both British imperialism and the rise of Islamic fundamentalism. This nostalgia for pre-revolutionary Egypt, when Cairo was indeed known as "Paris on the Nile", is common to many liberals in Egypt. A common understanding, not only referred to by liberals, is that Europe or the West in general contains certain highly commendable qualities with regard to governance and personal freedoms, but these qualities are betrayed by Europeans or Westerners in the way they conduct their foreign policy towards the Arab and Muslim world. Hilal writes that for a solution to "the Middle Eastern crisis" it is merely necessary "that America should honour the American values, most prominently the value of democracy".[38] This recalls the way in which Arab nationalists of the period before World War II attacked European colonialism without rejecting the project of cultural and political Europeanisation. As Hourani wrote of this disposition, "while the nationalists condemned British or French policy, the conclusion they drew was not that England or France were intrinsically bad but that they were being untrue to themselves. The appeal was to the "true" England and France, and the expectation was that sooner or later they would reassert themselves and understand that their interests were in harmony with those of the Arabs. [...] For the most part too, it was taken for granted that once independent the Arab states should adopt the characteristic institutions of European liberal society".[39]

---

36 Rida Hilal, *Al-Amraka wa al-Aslama: Ma'zaq 'Arab al-Yaum*, Al-Qahira, Dar Masr al-Mahrusa, 2004.

37 *Ibid.*, pp. 33-34.

38 *Ibid.*, p. 15.

39 Hourani, *Arabic Thought.*

The EU versus the US; the EU versus the Arab Union

In the Arab and wider Islamic world the United States of America first emerged independently in intellectual discussions at the end of World War I. America had been considered as little more than a cultural extension of Europe ever since the late nineteenth century, but the formulation by President Wilson of the Fourteen Points made for lively debates in Arab intellectual circles, as one of the points, the right to self-determination, was exactly what Arab nationalists were calling for. And so America entered the Arab imagination with positive connotations. Today, roles appear to have been reversed, as America plays the role of hegemon in the Arab world and Europe is considered the potential ally.

The European unification process is carefully followed in the Arab world for two reasons. Firstly, European unification is seen as an example for the Arabs. For almost an entire century Arab unity has remained a slogan more than a reality, while Europe has managed to unite within a time-span of little more than half a century. Secondly, as the EU grows stronger, the hope of many in the Arab world is that the EU will break the American hold over the region, force an equitable solution to the Palestinian-Israeli conflict, and serve as the economic centre with which the Arab world can trade profitably. These two points make for an image of Europe as one of admiration and hope for the future. This also means, however, that Europe is very often seen not in its own reality, but rather within the framework of a power struggle with America and the European position *vis-à-vis* the Arab world.

Figure 8.1. "Europe Launches the Greatest Plane in the World"

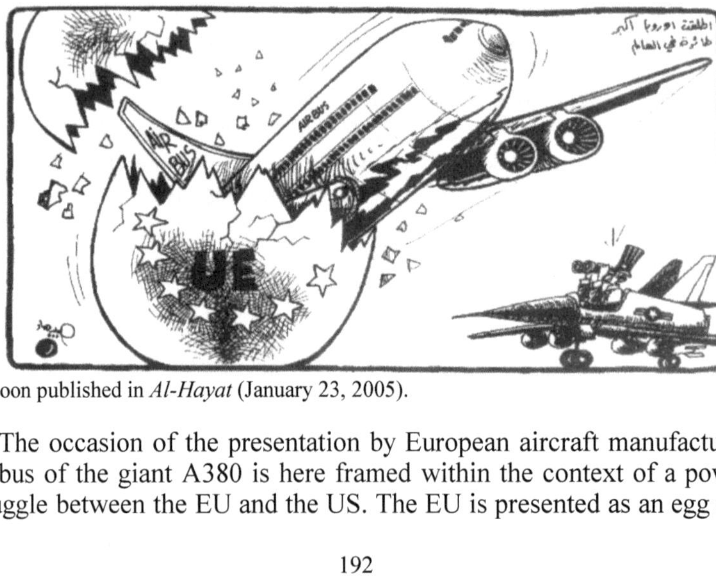

Cartoon published in *Al-Hayat* (January 23, 2005).

The occasion of the presentation by European aircraft manufacturer Airbus of the giant A380 is here framed within the context of a power struggle between the EU and the US. The EU is presented as an egg out of which suddenly emerges the majestic A380. The record-breaking civilian aircraft hovers over a startled Uncle Sam, sitting in a fighter plane. The spectator is thus presented with the image of a strong, youthful (witness the egg) and peaceful (civilian) Europe, as a counterweight to the contrasting image of an overpowered, aggressive (fighter-plane) America.

Even when the EU is discussed without focussing on its role as potential ally of the Arabs and counterweight against American power, we find that Europe can be used to say something about the political situation in the Arab world rather than about Europe itself. In a recent publication by the Lebanese political scientist Hasan Nafi, tellingly entitled *The European Union and the Lessons to be Drawn by the Arabs*,[40] the author's primary argument is that the EU's *sine qua non* is democracy. He argues that for each and every country which has joined and will join, the main condition has been and will be that the country is genuinely democratic. It is interesting to see that he argues that the whole European unification enterprise is a project concerned with safeguarding democracy, while in the European Union itself it is widely recognised that a major problem of the EU is the so-called "democratic deficit" between the peoples and their European political institutions. Unsurprisingly, the "lesson" that the book's title refers to lies herein: if the Arab world ever wants to unite, it must first democratise.

In a similar vein, a volume edited by the Egyptian veteran journalist Ibrahim Nafi, *What is happening in Eastern Europe?*,[41] aims at finding out how East European countries have changed their economies and political systems since their liberation from Soviet domination. In particular, the authors are on a quest to understand Eastern Europe's experiences with reforms towards political pluralism and an open market economy. Eventually, "its aim is to get to know the [...] lessons [to be drawn from these developments]" so that they may be "of benefit to us Egyptians and Arabs in planning for the future".[42] The contributions are not always mild in the way they compare the European and the Arab experiences with liberalisation (in the 1970s, under Sadat, Egypt embarked on economic liberalisation). A painful observation is made when it is stated that in reforming the economy, the East European countries have achieved more in ten years than Egypt in twenty-five.[43] In a contribution by Abdel Munim Said, the lesson is phrased as follows: "It is

40 Hasan Nafi', *Al-Ittihad al-Urubbi wa al-durus al-mustafada 'arabiyyan*, Bayrut, Markaz dirasat al-wahda al-'arabiyya, 2004.

41 Ibrahim Nafi (ed.), *Madha yagri fi Sharq Urubba?*, Al-Qahira, Al-Ahram 2001.

42 *Ibid.*, pp. 7-8.

43 *Ibid.*, p. 216.

clear that the dividing line between progress and retrogression [...] is the extent of the elites' agreement on the job ahead for the nation. [...] That said, what is there that the political elite in Egypt agree about?".[44]

Conclusions

In this chapter I have presented a wide range of images of Europe, hailing from the early nineteenth century through to our present day, allowing space for secular modernisers, religious reformers and fundamentalists, Marxists and nationalists, journalists and scientists. While this kaleidoscopic panorama is a useful approach if one seeks to dispel the idea that the Islamic world has a primordial adversity to infidel Europe, it makes it harder to formulate a concluding statement. The following points can nevertheless be made. The images of Europe as Other vary according to time, place and person, and they are often informed more by perceived needs of the self rather than by the reality of Europe. If Europeans read Al-Tahtawi's account of Paris today, they would probably feel a little embarrassed for the lavish praise heaped on the French capital and Europe. Similarly, many would be astounded to read how Muhammad Imara connects the Greek, Roman and Byzantine empires to the Western "imperialist invasion" of Iraq in our times. What Al-Tahtawi and Imara have in common is that their images of Europe are a vessel for their ideals. Tahtawi saw the need to modernise after the European model and therefore made an alluring image of Europe. Imara on the other hand seeks to alert his readership to the dangers of Western encroachment on the Islamic world, and therefore paints a picture of the West as being a sworn enemy of Islam. Similarly, when the process of European unification or the East European transformation to liberalism and pluralism is discussed, it is clear that the authors seek to formulate the way forward for the Arab world, rather than merely describing what has happened in (Eastern) Europe. Nevertheless, it would be very unwise to suggest that whatever people in the Arab or Muslim world think of Europe is essentially unrelated to the "reality" of Europe.[45] We have seen, for instance, that the politics of colonisation and the developments during and after World War I forced Arab intellectuals to restate their position *vis-à-vis* Europe. It was quite simply an undeniable fact that European ambitions in the Middle East frustrated the hopes and dreams of Egyptians and other Arabs who desired to build an independent nation. Similarly, the present conflict between Israel and the Arab

---

44 *Ibid.*, p. 225.

45 See Lewis, "Roots of Muslim Rage". A similar argument with regard to the supposed irrelevance of American foreign policy can be found in Margalit & Buruma, *Occidentalism*, p. 8.

countries plays an important role in the extent to which Arab regimes and intellectuals can comfortably ally themselves with the West.

When we review the images of Europe in the Arab world, we find that while there is no lack of generalist references to "the West", there is also a rather strong tendency to distinguish between Europe on the one hand, and America on the other. Referring back to the introduction to this chapter, we could say here that it should be acknowledged that in the Arab world there is more willingness to recognise the difference between *pars* (Europe) and *toto* (the West), than there is in Europe regarding the difference between Arab (*pars*) and Muslim (*toto*). In all likelihood, this difference lies in the image of Europe as a hoped-for ally of the Arab world. Perhaps that was also a strong ingredient in the row over the Danish cartoon drama. Had the cartoons been published in a newspaper in Patagonia, no one in the Arab world would have organised an objection for no one cares about Patagonia. Had the cartoons been published in America, it would merely have confirmed the convictions of many that America is against Islam. Perhaps only European newspapers desecrating Islam could have sparked the uproar of early 2006, exactly because of the expectation many have of Europe as a potential ally.

countries plays an important role in the extent to which Arab regimes and intellectuals can comfortably ally themselves with the West.

When we review the [illegible] of Europe in the Arab world, we find that while there [illegible]

CHAPTER 9

Latin America and Europe

Peter BEARDSELL

According to Jorge Luis Borges, the paradox of Argentina's history was "wishing to move away from Spain", but not from the whole of Europe. In order to cultivate a link with the past, he wrote, and in the absence of a tradition in their own country, Argentines turned to Europe, with whom they felt the link "in chronology and in the blood, very close to us".[1] It is hardly necessary to say that Argentina does not represent the whole of Latin America, but there is some justification for applying Borges' words more generally. The Peruvian sociologist, Aníbal Quijano, observed a "dualism in our intellectual manner", and argued that a "tension runs through everything and almost everybody in Latin America".[2] Latin Americans were not "exiled Europeans in these savage plains", he continued, but in their autonomous and distinct identity this dualism was an integral feature. In this chapter I will be analysing the ways in which Europe contributes to Latin Americans' identity.[3]

To generalise is risky, of course, especially when one is dealing with twenty different republics, and with people who descend not only from Europeans but also from indigenous civilisations.[4] When it comes to

1 Jorge Luis Borges, *Discusión*, Obras completas de Jorge Luis Borges 6, Buenos Aires, Emecé, 1961, pp. 158-60.

2 Aníbal Quijano, "Modernity, Identity, and Utopia in Latin America," *Boundary 2*, 20/3 (1993), pp. 140-155, quotations from p. 149.

3 This chapter is developed from three previous publications of mine: *Gods and Demons, Self and Other. Images of Europe in the Cultures of Latin America*, Hull, University of Hull Press, 1994 (Inaugural Lecture); *Europe and Latin America: the Identity of the Other*, Manchester, University of Manchester, 1995 (Manchester Spanish and Portuguese Working Papers 3); and especially *Europe and Latin America: Returning the Gaze*, Manchester, University of Manchester Press, 2000.

4 Although ongoing research suggests that Europeans reached the continent before Columbus, there are no certain records of that among indigenous peoples, and so for our purposes in this volume the first images of Europe date from the end of the fifteenth century. As far as nomenclature is concerned, my use of the term Latin America deserves some explanation. There are various names for the region and sub-regions, including South America, Spanish America, Iberian America, Central America, Middle America, Mesoamerica, various equivalents in Spanish, and the

images of Europe, there are sharply differing ethnic viewpoints, and in addition to these, the profound effects of hybridity. This proves to be the crux of the matter, since the hybridity implies not merely the mixture of blood but the blending of cultures. Images of Europe reflect this fusion, and the variations that occur among the twenty republics.[5] Social and political matters add to the complexity. A higher social class and a more conservative attitude to national and regional policies tend to coincide with a more positive image of Europe. Latin America's relations with the United States play their part too. Despite their parallel colonial pasts, these two regions have developed an unequal relationship, in which the northern neighbour's economic and military hegemony have led to a general Latin American antipathy. Europe has therefore offered an alternative with regard to both business and cultural ties. All of these factors have contributed to the content of this chapter. In particular, though, I shall focus on the contrasting images of Europe as invader, alien culture, and model of civilisation, before I address the more complex issues arising from Latin America's hybrid identity.

Europe as Invader

Accounts of the galleons that reached the Gulf of Mexico in 1519 provide the first images inspired by visitors from Europe. One report described a "mountain range or small mountain floating in the midst of the water", a second "two towers or small mountains floating on the waves of the sea".[6] The strangeness and sense of great size that dominated these impressions of eyewitnesses were to become features of subsequent visions too. Montezuma's messengers confirmed to him that "it is true that strange people have come to the shores of the great sea [...] They have very light skin, much lighter than ours. They all have long beards, and their hair comes only to their ears" (p. 17). The messengers were greatly amazed by the extensive use of iron: "They dress in iron and wear iron casques on their heads. Their swords are iron; their

autochthonous term *Abya-Yala* (meaning "land in complete maturity"). But they are all problematic in some way or another, and several even have neo-colonial overtones. Latin America obviously has its own contradictions, but it has one great advantage: generally speaking, it is the name preferred and accepted by most Latin Americans.

5 Eighteen republics have Spanish as their official language; Brazil has Portuguese; Haiti French. But Indian languages, such as Quechua and Aymara, have a recognised legal status in some countries. Peru, for example, has two official languages: Quechua and Spanish.

6 Miguel León-Portilla (ed.), *The Broken Spears. The Aztec Account of the Conquest of Mexico*, English translation by L. Kemp, Boston, Beacon Press, 1992, p. 16. Page references to this book are given in brackets in the text.

bows are iron; their shields are iron; their spears are iron". Also difficult to explain were the Spaniards' horses, which, being unknown in the Americas, looked like deer "as tall as the roof of a house" (p. 30). As for the guns, their sound and power terrified the witnesses: "A thing like a ball of stone comes out of its entrails [...] shooting sparks. The smoke that comes out with it has a pestilent odor [...] If the cannon is aimed against a mountain the mountain splits and cracks open" (p. 30). This image of fearsome power was enhanced by the Spaniards' dogs, which the messengers saw as "enormous", "tireless and very powerful", with eyes that "flash fire and shoot off sparks" (p. 31). Battles with some of the coastal Indians demonstrated the military prowess of these strangers, and Montezuma's messengers were treated in a way that ensured the newcomers' aggressive and assertive nature would be reported.

Figure 9.1. "Extract from Codex Ríos, 1 Reed"

Facsimile of Toltec annals of the period 1516-25.[7]

It is useful to see how the first encounters between Europeans and the people of Mesoamerica were represented in pictographic writing found in a Toltec script recording the period 1516-25. Figure 9.1 reproduces the appropriate section of this script, where the Indians record European might and hostility. There is clearly a sharp contrast between the European on the left, mounted on a horse with a sword held aloft, and the Indian figure on the right, standing before him with an offering of fruit. Dominance, aggression, and military threat are not all that the

[7] Figure 9.1. is an edited extract from Codex Ríos, 1 Reed. The Codex Ríos is held in the Vatican Library. A facsimile of the section of the screenfold containing 1 Reed is on p. 87v in R. Wauchope (ed.), *Handbook of Middle American Indians*, Austin, University of Texas Press, 1964-75, Vols 14 and 15, No. 270; and is also reproduced in Gordon Brotherston, *Image of the New World*, London, Thames and Hudson, 1979, p. 31.

image suggests about the European. The shield is topped by a cross with a trailing banner, thrust towards the Indian as though to demand acceptance of the religious ideas of an emissary from another land. In other words, the Aztecs represented themselves as people humbly offering a welcome to the strangers, but chose to depict the newcomers as militaristic aggressors seeking to impose their beliefs. Of course there was a deliberate political bias in this extremely selective representation, which disguises the ambiguity of the "welcome" offered to Hernán Cortés, but for our purposes it illustrates – albeit in an exaggerated fashion – the early image of Europeans recorded in the annals of the conquered people.

For the Aztec ruler and his advisers the strangeness, power, and ruthless determination of the newcomers were compounded by the timing of their arrival and the situation in which they were first seen. The Europeans landed just when omens had been predicting some kind of catastrophic event. Moctezuma imagined this might be the return of Quetzalcóatl, a mythical ruler-cum-deity who – according to tradition – had sailed into banishment or chosen exile in the Gulf of Mexico, and was expected to come back one day to claim his rightful position. With their white skin and beards, unfamiliar apparel, purposeful advance, and apparent invulnerability, the Europeans could have been gods, and the place from which they came, beyond the sea, could have been a mythical home of the deities. In this aspect, then, Europe was in effect accommodated to comply with the Indian belief system.

Not surprisingly, the image of Europeans as gods did not last long. Human traits soon surfaced, especially greed for gold, voracious eating habits, cruelty, and vulnerability to injury and illness. The behaviour of the Europeans was then compounded by three centuries of colonial rule, which destroyed much of the Indians' old order and caused general distress and suffering. Europe's image assumed characteristics determined by these experiences. It was the provenance of ruthless invaders, the imposer of a new culture, and the perpetrator of social injustice.

These features are strikingly illustrated in the pages of a "letter" for the king in Europe written during the period 1585-1615 by an Inca prince.[8] Many of these pages bear drawings with essentially negative images of the Europeans. Some depict outrages committed during the Spanish conquest, such as the burning of houses and the people inside

8 Felipe Guaman Poma de Ayala, *El primer nueva corónica y buen gobierno*, edited and translated by C. Dilke, *Letter to a King. A Picture-History of the Inca Civilisation by Huaman Poma*, London, George Allen & Unwin, 1978. The "letter" (which was never delivered to the king) is a book of over one thousand pages composed during the years 1585-1615. Around four hundred pages consist of line drawings.

(on the page headed "Conquista don Francisco Pizarro"). In general the Europeans are drawn carrying swords, lances, whips, daggers and sticks, while they remonstrate, command, control, and demand gold. The injustices perpetrated during the colonial period are the subject of many other pages.

Figure 9.2. Felipe Guaman Poma de Ayala, "Papal World"

"Pontifical mundo", in *El primer nueva corónica y buen gobierno* (1585-1615).

One of them, "Pontifical mundo" (Papal World), gives us a remarkable idea of Guaman Poma's image of Europe (Figure 9.2). In the lower section of the drawing Europe is seen as a land of town squares and churches ruled by the Pope from the Centre, named as Castille. Above the line, to draw a neat comparison, the author depicts the order of the Inca world, ruled from its Centre, Cuzco – an order destroyed by the conquest. Since the drawing obviously seeks to demonstrate a crude similarity in cultural, religious and political systems, it also suggests the idea that the relationship between Europe and the Indian world should be essentially of equals. It is not difficult to see that Europe's imperial dominance is the issue here.

Europe as Alien

Memory of the trauma caused by the disastrous collapse of the old systems was reinforced by the perception of deep social inequalities and the experience of extreme poverty and harsh treatment. This legacy of the Conquest is reflected even today in cultural traditions observed in rural districts with a high proportion of indigenous inhabitants.[9] Every year, for example, certain festivals include a public spectacle re-enacting the Conquest in the form of a play, mime or dance (usually known generically as the "dance of the Conquest"). While the specific content varies from place to place, researchers have found that the narratives have one or two significant and fundamental differences from the events recorded by official histories. In one Peruvian version the King of Spain, presented with the head of the Inca chieftain Atahualpa, does not allow Pizarro to enjoy his victory but censures him and promises him punishment. In one of several Mexican versions the outcome of the Conquest of Mexico is reversed, presenting Moctezuma as the victor and Cortés pleading for forgiveness. Many other variations are found, but they all have in common the need to alter the memory of events that proved disastrous and unjust. It follows from this that despite the region's liberation from Spanish colonial rule during the nineteenth century (most of the wars of Independence achieved their ends between 1810 and the 1830s), the sense of Europe as the seat of an alien power was not completely eradicated from the popular culture of many Latin American republics (in particular those with a substantial Indian population). There is a perception that, although the alien power is nowadays based in the cities of Latin America or in the United States, those who hold it are essentially identified with colonial Europe.

After independence from Spain there was a fundamental split between wealthier, more conservative Latin Americans on the one hand, and the poorer economic sectors of society on the other. The wealthy, who often had blood ties with Europe, encouraged the eagerness of Britain, Germany and other European countries to fill the economic and political vacuum. The poor, who tended to include most people of indigenous origin, saw advantages in maintaining a distance from Europe. Such class and ethnic factors often determine Latin American attitudes to Europe to this day. Intellectuals, aligning themselves with the oppressed people, have expressed their sense of alienation from capitalist society, which they tend to identify with urban areas, the

9 See William Rowe & Vivian Schelling, *Memory and Modernity. Popular Culture in Latin America*, London & New York, Verso, 1991; and N. Wachtel, *La Vision des vaincus. Les Indiens du Pérou devant la conquête espagnole 1530-1570*, Paris, Gallimard, 1971.

United States, and Europe. An example worth noting is the Peruvian poet Antonio Cisneros, whose collection *Like a Fig Tree on a Golf Course* (1972) contains poems reflecting his personal experiences while living in Europe, in particular his lack of adjustment to the foreign culture. A revealing image is the pervasive war memorials, reminders of conflicts that he never knew, "triumphal arches which celebrate my condition as a slave, as the son of men who eat rice".[10] His sense that the colonial spirit has not died inspired the poem "Denunciation of Elephants (too well thought of recently)", which ironised the contradiction between commercial exploitation and the natural world. Elephants epitomised the way Africa had to adapt to the will of foreign powers. They had learned English, paraded in front of visitors in city zoos, and stood in the background while the European on safari sipped a cup of tea.

Many Latin Americans have been alienated by Europe's unnatural environment, especially the industrial landscapes (which stand in contrast to the vast areas of wild nature in Latin America). An Ecuadorian poet stranded in Europe in 1928, Jorge Carrera Andrade, wrote poems as bulletins to send home. In them he greeted Ecuadorians, who lived close to the land, from what he called "these manufactured landscapes", with towers, cranes, mechanical ploughs, and chimneys.[11] Decades later his vision had not changed. The imprisoning effect of a city was likened to a cage. Trees and grass confined within railings was a repellent sight, contrary to the natural order, and it struck him as being like enclosing man's natural spirit. In "Keep off the Grass" he rebelled against this unnatural culture: "I know your world of imprisoned parks, of birds that sell all the music of the woodland in a cage" (p. 433). Parks where trees were arranged in neat rows offered an image of European children, trained to behave unnaturally. Civilisation was denying itself the essential requirements for life, and in consequence was falling into a state of decay. This idea gave him the image of Europe as a withering tree: "Europe, you are a chestnut tree with leaves about to break away and die: stretch out your yellowing hand to me before it flies away" (p. 433).

Europe as Civilisation

The complexities of Latin America's social, political and ethnic characteristics since independence mean that the essentially negative images of Europe presented in the previous section are also counterbal-

---

10 Antonio Cisneros, *Como higuera en un campo de golf*, Lima, Instituto Nacional de Cultura, 1972, p. 133. Translations are mine.

11 Jorge Carrera Andrade, *Obra poética completa*, Quito, Casa de la Cultura Ecuatoriana, 1976. Page references to this edition are given in the text. Translations are mine.

anced by positive images. To some extent this was a consequence of economic trends encouraged by European dominance, particularly the *régime* of *laisser-faire*, in which commerce, foreign investment, and the presence of foreign companies and personnel created what a prominent Brazilian economic analyst called a "mercantile bourgeoisie, liberal in outlook, progressive in the sense that it was 'Europeanising'".[12] This led, naturally enough, to a certain identification of Europe with Latin America's ruling class of landowners and merchants – an image that the influx of large numbers of relatively poor European immigrants late in the nineteenth century did not entirely dispel.

As the new republics explored their identities during the nineteenth century, then, certain influential sectors of society continued to regard Europe as a source of inspiration for the development of their own culture. It is true that there was a search for self-determination and a strong sense of nationalism, but parallel with this there always remained a powerful Europhile strain. One of the ideas on which some politicians and intellectuals based their programmes for development was that Latin America's future depended on the struggle between the barbaric forces prevalent in untamed nature, and the civilising influences introduced as population centres spread throughout the region. This argument was most famously expressed by an Argentine politician, Domingo F. Sarmiento, in 1845, during his years in exile before assuming the presidency of his country. Having witnessed a period of notoriously savage dictatorship in Argentina under Juan Manuel Rosas, a landowner whose support was drawn partly from gauchos, Sarmiento reasoned that Rosas represented the triumph of barbarism. Moreover, since the land itself was essentially hostile to civilisation, the solution for Argentina was to extend the influence of the cities over the land and thereby increase the impact of European values, ideas and ways of life.[13] It was on these principles that the foundations of modern Argentina were laid during the two decades from 1860 to 1880. An important strategy was that of simultaneously increasing the population of the country and the cultural impact of Europe by means of immigration, ensuring that Argentina should be peopled – as one of its historians has remarked – by "races that might improve the ethnic level and the mentality of Argentine creoles".[14]

12 Celso Furtado, *Economic Development of Latin America. Historical Background and Contemporary Problems*, Cambridge, Cambridge UP, 1976, pp. 35-36.

13 Domingo F. Sarmiento, *Facundo o la civilización y la barbarie*, 1845.

14 F. Luna, *Breve historia de los argentinos*, 14th edition, Buenos Aires, Planeta, 1997, p. 122. (My translation of his words, "Razas que puedan mejorar el nivel étnico y la mentalidad de los argentinos criollos").

Figure 9.3. Juan Cordero (1824-1884), "Columbus at the Court of the Catholic Monarchs", 1850

Cristóbal Colón en la Corte de los Reyes Católicos (1850), oil on canvas, 181x242.5 cm, Museo Nacional de Arte, CONACULTA, INBA, Mexico.

In the context of such a strategy, it is not surprising that Europe's self-image as the centre of culture in an expanding area of enlightenment should be adopted by certain political groups throughout Latin America. The concept is well illustrated by a Mexican painting of the middle of the nineteenth century, Columbus at the Court of the Catholic Monarchs (1850), by Juan Cordero (Figure 9.3). This depiction of Columbus appearing before King Ferdinand and Queen Isabella with evidence of his discoveries emphasises ideological implications of the event. The centre of power is evident in the central position occupied by the monarchs, appropriately elevated on a platform. A circle of bright light draws the eye of the observer to the royal figures in the centre. As Columbus stands submissively before them, he gestures towards the edge of the painting, the palm of his hand turned upwards in an act that is simultaneously one of pointing and offering. And at the very edge of the composition (on the periphery of the event) stoop the people that Columbus is presenting for inspection. They are barely perceptible in the shadows, as though the darkness were an image of the unknown lands from which they come. Here is an image of the Indian people being brought to the seat of power, led out of obscurity into the light of knowledge.

In fact, the image of Europe as the centre and origin of things permeated – and to some extent, one might say, permeates – Latin American culture at all levels. It has even produced a peculiar slant to a universal myth: in some families children are told that babies come into the world carried by storks from Paris. In wealthy circles a visit to Europe has always been desirable for Latin Americans with political, military, artistic, and commercial aspirations. Indeed, during the nineteenth century it became "one of the most important means by which liberal values and ideas were transmitted to the countries of their birth during the nineteenth century".[15] Those who could afford it took holidays in Europe, and sent their sons and daughters to be educated there.

At the time of independence from Spain, the most widely known foreign language was French. Writers were especially susceptible to the notion that Paris was the cultural centre *par excellence*, and although they assimilated ideas and modes under various European influences, they learned especially from French intellectuals. Some of them, inspired by the French Parnassians, Symbolists and Decadents during the last decades of the century, developed new trends and took them from Latin America back to the former imperial centre, helping to regenerate literature in the Spanish language. Although the twentieth century often witnessed a strong reaction against this Europeanising trend – sometimes denigrated as servile imitation – there was essentially no end to the search for inspiration and models in Europe. The age of dependence had closed, and the crux of imitation occurred at the time of the avant-garde, but the concept of Europe as a source of cultural continuity and a model of literary sophistication persisted. For writers as diverse as Guillermo Cabrera Infante (Cuba), Isabel Allende (Chile), Ariel Dorfman (Chile) and Mario Vargas Llosa (Peru), political exile in European capitals played its part in this process.

A slightly ambiguous expression of the image of Europe in Latin America is the variety of attitudes towards the annual commemoration of Columbus' arrival on 12 October 1492. This day, celebrated in the United States as Columbus Day since 1792, was first introduced into the calendar of festivals in Latin American republics after its inauguration in Argentina in 1917. It is known as *El Día de la Raza* (The Day of the Race) or *El Día de la Hispanidad* (The Day of Hispanic Culture), terms that represent its intention: not only to remember the bond with Europe but also to celebrate Hispanic roots and emphasise Latin as opposed to Anglo-Saxon identity. The potentially controversial nature of the festival

---

15 Rory Miller, Introduction to P. Deche and C. Diamond, *Latin Americans in London: a Select List of Prominent Latin Americans in London, 1800-1996*, London, ILAS, University of London, 1998, p. 1.

is immediately visible in words of the Argentine government's original decree, issued on 4 October 1917.[16] Article 1 declares the discovery of the Americas to be "the most momentous accomplishment achieved by humanity throughout the ages"; Article 2 attributes this feat to the Hispanic character; and Article 3 praises it all as "the miracle of conquering for civilisation the immense estate now flourishing among nations". Not surprisingly, the hyperbolic tone does not accord with everyone's sentiment. Apart from that, the positive attitude to the conquest and its supposed benefits flies in the face of the views held by many, in particular indigenous people and left-wing political groups. While a full week of activities is not uncommon in certain countries, such as Argentina (where Italian as well as Hispanic ties make this a significantly European event), a low-key approach to 12 October is normal in others, notably Mexico, where the most likely form of acknowledgment is a demonstration by the indigenous movement (sometimes accompanied by assaults on the statue of Columbus in the Avenida de la Reforma). In fact, the word "Raza" allows indigenous people to celebrate their own roots on the day – in Argentina as well as Mexico. For our purposes here, the important point is not so much that 12 October may inspire both positive and negative attitudes, but that it signifies a claim to identity. In the official calendar of Latin American countries *El Día de la Raza* is a reminder European origins for many families, and an image of European historical connections for everyone.

Europe and Identity

In previous sections of this chapter I have attempted to show how Europe's image in Latin America varies from one sector of society to another. In practice, of course, there is no clear-cut distinction, just as there is no complete segregation: social and ethnic groups overlap and merge. Essentially, Latin America is an area where any dualism yields to cultural hybridity in a postcolonial context. It is useful here to note that in an analysis of the effects of colonial power in his book *The Location of Culture*, Homi Bhabha sees hybridity as the key to the "strategic reversal of the process of domination", the "revaluation of the assumption of colonial identity". Colonialism seeks to produce "the silent repression of native traditions", but causes instead a mixture of traditions. In that mixture lies the basis for the colonised people's assertion of their own identity.[17] Bhabha develops the idea into an argument that

[16] A copy of this document is stored in the file "El Día de la Raza" in the archives of *La Nación*, Buenos Aires.

[17] Homi Bhabha, *The Location of Culture*, London & New York, Routledge, 1994, p. 112.

the "subaltern" is enabled to reverse the cultural domination of the colonial power by "turning the gaze of the discriminated back upon the eye of power". It is important to emphasise that Europe's image in Latin America is – partly, at least – the consequence of returning the gaze in this fashion. This process, in turn, is a feature of Latin Americans' search for their own identity.

Bhabha's point is implicit in the distinction that anthropologists often make between acculturation, where the indigenous cultures are replaced by European ones, and transculturation, in which the transformation of cultures is mutual.[18] It is the second process that concerns me here. In his study of transculturation in Latin America, the Uruguayan cultural critic, Angel Rama, pointed out that while the spread of cities has the effect of extending the culture of the metropolis throughout Latin America's regions, a counter-process can often be found, in which contributions of modernity are adopted and regional cultures are revised in their light without losing the native heritage.[19] It is useful to follow Rama's line of argument, replacing the city and the regions in this model with Europe and Latin America. In this light, we can see that hybridity involves a partly conscious policy by means of which Latin Americans take hold of selected European cultural developments and use them in revising their own cultural traditions. Some of the forms taken by transculturation therefore amount to a strategy of adapting while adopting.[20]

In this light a visit to Europe may be seen as more significant than a mere cultural tradition or educational pilgrimage. It can be a means of stepping out of oneself in order to adopt the Other's viewpoint. While on the other (European) side of the Atlantic Ocean, Latin Americans – and especially Latin American intellectuals and artists – can become aware of how they are seen by Europe and incorporate that view into their response. They are often frank about the benefits. The Colombian novelist Eduardo Caballero Calderón remarked that from Europe the Latin American may see his own continent with a historical perspective. The Argentine novelist Eduardo Mallea admitted that he needed the

18 Rowe and Schelling, *Memory and Modernity*, p. 18.

19 Angel Rama, *Transculturación narrativa en América Latina*, Montevideo, Arca, 1989, p. 29.

20 This is more or less what Andrea Giunta calls "appropriation of appropriation" in her essay "Strategies of Modernity in Latin America", in G. Mosquera (ed.), *Beyond the Fantastic. Contemporary Art Criticism from Latin America*, London, Institute of International Visual Arts, 1995, pp. 53-66. To be precise, what she means by this is the effect achieved when Latin Americans repeat something that Europeans have already appropriated from Latin America. They know that they are being treated as Europe's Other, but by repeating the European act they present the Otherness as an inherent aspect of their identity.

distance in order to see Latin America positively. The Mexican painter David Siqueiros explained that he and his colleagues took an exhibition to Europe "in order to find out what the European critics thought of us [...] It would give us a means of evaluating our own contribution".[21] Even without travelling to Europe, a familiarity with European writers and artists can permit the same kind of insight. In other words, Latin American writers and artists can regard European literary and artistic treatments of their continent as a means of learning more about themselves. Or, to put it another way, Latin Americans can assimilate, or – ultimately – exploit, Europe's own image of Latin America.

A famous example of the European image of Latin America is the nineteenth-century myth according to which people living on the American continent were closer to nature. A European writer popular with Latin American readers during the nineteenth century, Francois-René de Chateaubriand, expressed the European Romantic attitude to America in his novelette *René* in the following words: "Europeans, endlessly anxious, are compelled to construct solitudes for themselves. The more our hearts are filled with noise and tumult, the more we are attracted by silence and tranquillity. Oh happy savages! If only I could enjoy the peace that is always with you! [...] you, quietly sitting beneath your oak trees, and letting the days flow by without counting them".[22] This idealistic image of Latin America was in effect one aspect of its more general role of dream-fulfilment. According to Eduardo Caballero Calderón, Europeans tended to project Utopian and humanist dreams on a continent that they considered a blank page; a continent that was "the subconscious of Europeans; a Europe unrestrained and unshackled, free of prejudices, violent, chaotic, charged with a heavy burden of unsuccessful acts and unsatisfied desires".[23]

In Latin America the effects of this European view had a tremendous impact, especially during the course of the nineteenth century. In the growing urban centres, it helped to determine the attitude that the metropolis held towards the interior, while in the regions themselves writers subsumed it within their portrayal of the wild areas of their countries, including their landscapes, climates and inhabitants. A brief example will illustrate this point. Horacio Quiroga, a Uruguayan writer who spent a good deal of his life as a settler and pioneer in the Misiones

21 David A. Siqueiros, *Art and Revolution*, translated by S. Calles, London, Lawrence & Wishart, 1975, pp. 145, 151.

22 Francois-René de Chateaubriand, *Le Génie du Christianisme*, 1802. Quoted from P. Crump (ed.), *Atala, René*, Manchester, Manchester UP, 1961, pp. 86, 90 (my translation).

23 Eduardo Caballero Calderón, *Americanos y europeos*, Madrid, Ediciones Guadarrama, 1957, pp. 42-43 (my translation).

region of Argentina, composed short stories for the readers of magazines in Buenos Aires in the early twentieth century. It is noticeable that many of Quiroga's pioneering protagonists are of non-Hispanic European descent. One of his collections, *Los desterrados* ("The Exiles", 1926), has immigration and assimilation as underlying themes. The story "The Immigrants", in the collection *El salvaje* ("The Savage", 1920), has these as the principal themes, and shows clearly how Quiroga exploits the mutual images that Europeans and Latin Americans have of each other.

In essence, "The Immigrants" tells of a man and a pregnant woman who die of exhaustion from travelling on foot in stormy heat through wild country. But the key to an appreciation of this short story lies in the few details that the author provides about the background of his characters. They have come from Europe with a dream of making money and returning happy and rich.[24] With the need to travel across difficult terrain in the extreme heat, the woman's pregnant condition – which at first had been a symbol of a new start for the couple – proves fatal. They had arrived in that land with hope, but the man's attitude has changed completely, "his whole mission concentrated on a single obstinate idea: to wrest the adored body of his wife from that wild and hostile country" (p. 42). In other words, Quiroga is illustrating the contradiction between the concept of untamed Latin America as a place in which Europeans may fulfil their dreams, on the one hand, and the reality of the region as an autonomous entity indifferent to extraneous designs, desires and needs. The author is aware of the European myth of Latin America as a land of opportunity, and a land where human beings may live in harmony with nature, and has used it as a foil in order to demonstrate the reality as opposed to the myth. Latin American nature has its own character, which is capable of extremes, is not easily malleable, demands respect and poses dangers. Human beings may adapt and survive – even succeed, as he shows in some of his stories – but they may also fail or perish – as he shows in the majority. He has used the European myth precisely because in Buenos Aires his readers will have been influenced by it. Europe's image becomes a literary device.

When Gabriel García Márquez delivered his speech on receiving the Nobel Prize for literature in 1982, he too made use of Europe's image among Latin Americans. Playing on the world's familiarity with his novel *Cien años de soledad* (One Hundred Years of Solitude, 1967) he claimed that solitude was the source of Latin America's problems, solitude caused by Europe's (and the world's) lack of understanding and

---

[24] Horacio Quiroga, *El salvaje y otros cuentos*, 3rd edition, Buenos Aires, Losada, 1976, pp. 41-42.

solidarity. Europe (and the rest of the world) could help Latin America to free itself, creating a Utopia that would be the opposite of the present reality. He was remembering, of course, that Europe had imagined Latin America as a location for its Utopian ideal. Amerigo Vespucci's *Mundus novus* (1503) and Thomas More's *Utopia* (1516) were among the influential texts, and besides, in a general sense, it was clear that at the time of the *conquistadores* Europe was disenchanted with its old system and was searching for an opportunity to realise its ideals. In 1992 the Mexican author, Carlos Fuentes, even went so far as to say that America was *invented* by Europeans to meet their imagination and desire.[25] An interpretation of the end of *Cien años de soledad* in the light of his Nobel speech shows how the political implications for Europe and Latin America are carefully incorporated. The century of solitude, rounded off by the extermination of the village, Macondo, may be seen as a consequence of ideals being corrupted by faulty human behaviour, in particular the tendency of characters to isolate themselves and to look inwards. The last words of the novel indicate that the Buendías are a race condemned by their solitude, and that they will have no second opportunity to escape it. But the Nobel speech turns these words on their head. If Europe (and the rest of the world) were to help Latin America to free itself it would be possible to envisage the continent as a Utopia, where love was true, where happiness was possible, "and where races condemned to a hundred years of solitude may have at last and for ever a second opportunity on earth".[26] With these words García Márquez turned Europeans' image of Latin America to political advantage. In the novel itself there is ample evidence that the European perspective has been appropriated and ironically undermined (especially through the system of religious belief expressed through the Bible), so that the reality of Latin America emerges as a contradiction to the concept, ideal, Utopia or vision.

This process of what we might call *critical appropriation* is one, therefore, in which Europe's image is that of an entity viewing Latin America as its Other, and in which Latin Americans exploit their objectification to their own advantage. It may be found, in fact, at the very heart of the mode that, in the late twentieth century, became most closely associated with Latin America's international literary fame: magical realism. If realism is "a mode of writing that gives the impres-

[25] Carlos Fuentes, *El espejo enterrado*, Mexico, Fondo de Cultura Económica, 1992, p. 133.

[26] Gabriel García Márquez, "La soledad de la América Latina", *Casa de las Américas*, 23, No. 137 (1983), p. 5 (my translation).

sion of recording or 'reflecting' faithfully an actual way of life",[27] or "an overall faithfulness in the rendition of actuality,"[28] then magical realism is essentially these same things *plus* (or in spite of) the presence of elements that appear to defy reason. It may include fantasy, imagination, dream and mythology, but the ultimate assumption is that the work of art is attempting to depict a phenomenon that exists outside the artist's (and the receptor's) imagination as well as within it. García Márquez once explained that, since the time of the Europeans' discovery of Latin America, they had often been amazed by what they saw there. Their wonder had even caused them to confuse reality with imagination, legend and myth. He, on the other hand, was a creature of that "outrageous reality", and "[has] had to make very little use of the imagination". For Latin American writers like him, "the greatest challenge [...] has been the inadequacy of conventional means to make our life believable".[29] Magical realism, then, is not strictly the product of imagination (and therefore distinct from "fantastic literature" or "artistic realism") but the search for a means of making Latin American life believable to outsiders like Europeans, who do not share their experience.

Convincing explorations of the origins of magical realism have demonstrated that important elements may be identified in European art and literature from around 1920. In Germany the term was used in the book *Nachexpressionismus. Magischer Realismus* by the art critic Franz Roh in 1925, while the first use of the term in Spanish was in the book's Spanish translation, *Realismo mágico. Post-expresionismo*, published in Madrid a year later.[30] In Latin America the term was not used until 1948, though by the 1960s a prolific discussion of magical realism in Latin American literature served to associate it with the *Boom Latinoamericano*, which brought international fame. Meanwhile, it is important to note that surrealism (the impact of which on magical realism is profound) had exerted a great influence in Latin America, since its emergence during the avant-garde period in Europe. There is little doubt, therefore, that magical realism owes its origins to European artistic and literary modes. But it is not difficult to see that the popularity of Latin American fiction during the Boom period (at its height between 1962

---

27 Chris Baldick, *The Concise Oxford Dictionary of Literary Terms*, Oxford & New York, Oxford UP, 1990, p. 184.

28 Joseph T. Shipley (ed.), *Dictionary of World Literary Terms*, London, George Allen & Unwin, 1970, p. 266.

29 "La soledad de la América Latina", p. 4.

30 See A. Márquez Rodríguez, "Deslinde entre el *realismo mágico* y *lo real maravilloso* a propósito de la novelística de García Márquez", in Ana María Hernández de López (ed.), *En la punta de la Mira: Gabriel García Márquez*, Madrid, Editorial Pliegos, 1985, p. 338.

and 1972) played a significant part in the frequency of the term's use in Europe and North America. In fact, there are signs from around the 1960s of a growing tendency to apply it retrospectively to European works. What this suggests, therefore, is that the term magical realism was found to be so apposite in Latin America that it was appropriated by writers there, modified and enriched by an environment to which it was naturally suited, and then exported to other parts of the world, including its place of origin in Europe.

Finally, it is important to add that Europe's image can be symbolic for Latin Americans while they are exploring their own identity. By this I mean that Europe can be almost abstracted, and converted into a myth or even a metaphor representing the objective of a quest. Let us recall that some writers have felt alienated by Europe's industrial scenery, and repelled by its unnatural environment, while others, more positively, have viewed Europe as a model of civilisation, or a location associated with their origins. Genuine as the feelings are, it is not difficult to perceive that at times the writers are essentially more interested in working out their personal problems than in commenting on Europe. As Antonio Cisneros examined his sense of alienation and exclusion while living in Europe, he also discovered an illogical inability to return to Peru. It became a matter of confusion over his identity: "I have to fly to Lima, though I'm afraid of not recognising myself between the photo/from my family photo" (p. 132, the confusion is represented by the hesitation over the words "between" and "from" – in Spanish "entre la foto/de mi foto en familia".) Alienated from Peruvian society too, he was voicing his experience of estrangement from modern society, and ultimately his metaphysical anguish at not belonging in this world. Though he was at times a poet of social protest, he was also a poet of individual anguish when Britain and Peru, Europe and Latin America became ultimately metaphors by means of which he expressed facets of his whole vision of the world.

A completely different example of Europe's metaphorical function may be found in Julio Cortázar's Argentine novel *Rayuela* (Hopscotch) (1963), described by one critic as "the cosmopolitan novel of the 1960s" and said by another to be "inscribed against Western society".[31] It is not difficult to see the principal metaphors, since they are integrated in the novel's structure. The first part takes place in Europe (Paris), and the

[31] Steven Boldy, "Julio Cortázar, *Rayuela*", in Philip Swanson (ed.), *Landmarks in Modern Latin American Fiction*, London and New York, Routledge, 1990, p. 118; and Julie Jones, *A Common Place. The Representation of Paris in Spanish American Fiction*, Lewisburg, Bucknell UP, 1998, p. 25. Page references for the novel, given within brackets in the text, are for the following edition: Julio Cortázar, *Rayuela*, Buenos Aires & Mexico City, Alfaguara, 1966.

second in Latin America (Buenos Aires); the first is headed by the words "Over there on that side", the second "Over here on this side". The Argentine protagonist, Horacio Oliveira, who admits that "searching was my trait" (p. 26), is apparently living out some kind of spiritual journey or quest for meaning. At the end of each part the motif of the game of hopscotch occurs conspicuously, reinforcing the sense that in his journeys Oliveira is engaged in a symbolic attempt to reach (by kicking the stone) the zone at the top of the hopscotch, which the narrator calls "Sky" or "Heaven" ("Cielo"). In this novel, where metaphors are carefully contrived, boldly proclaimed, extremely self-conscious and, above all, self-referential, Paris is openly named as a metaphor (pp. 154 and 205). The connection Europe-Latin America becomes one aspect of many interlinking (and often contrasting) concepts in Oliveira's mind: "the moral, the immoral and the amoral, justice, charity, things European and things Latin American, day and night […]" (p. 100). Linguistic terminology reminiscent of psychoanalysis (Freud, Jung and Lacan) permeates the meditations: Oliveira is worried by "the weight of the subject in the notion of the object" (p. 36), perplexed that "otherness lasts with us as long as a woman does" (p. 119), and troubled by a sense of split personality, with one self walking in the Latin quarter of Paris "while the other, the other […]" (p. 136). It all signals the idea that in crossing from one side of the Atlantic Ocean to the other, Oliveira makes the individual's journey in which the Latin American Self searches for its Other. Europe (Paris) therefore serves as the alternative location, the Other.

In this chapter I have chosen to simplify an already complex matter by referring to Europe as a single entity. Naturally enough, Latin Americans do not always lump together the various European nations in this way. For some, the Conquest and imperial power means Spain, while for others, Spain is the mother country and a favoured trading partner. France (and particularly Paris) tends to represent the centre of culture. Britain signifies the nineteenth-century economic and imperialist power (especially for Argentina), and other nations have images derived from their specific contributions to immigration, farming, exports, technology, military training, etc. In spite of this, there is no doubt that, just as the United States has an image in Latin America, so has Europe as a whole, and not simply because of the modern entity created by the European Union. Europe is both distant and yet within reach (the Atlantic Ocean being reduced in popular parlance to "el charco" (the pond). It is both alien and familiar, both Other and Self. In searching for an individual identity, each Latin American republic looks at its natural environment, its regional history, its indigenous people, and its distinct characteristics. But at some time or other, when political circumstances are right, it also turns to its European heritage. Even today, in the early twenty-first century, with nationalist and anti-imperialist movements predominant, Europe plays a fundamental role in Latin America's sense of identity.

References

Aaron, D., *The Americanist*, Ann Arbor, University of Michigan Press, 2007.

Abdo, G., *No God but God. Egypt and the Triumph of Islam*, Oxford, Oxford UP, 2000.

Abduh, M., *Risalah al-tawhid*, Al-Qahira, 1953.

Abu-Rabi`, I.M., *Contemporary Arab Thought: Studies in Post-1967 Arab Intellectual Thought*, London, Pluto Press, 2004.

Adler, K., "'We'll Always Have Paris': Paris as Training Ground and Proving Ground", in K. Adler, E.E. Hirshler, H.B.Weinberg, *Americans in Paris, 1860-1900*, London, National Gallery, 2006.

Agelopoulos, G., "Perceptions, Construction, and Definition of Greek National Identity in Late Nineteenth-Early Twentieth Century Macedonia", *Balkan Studies*, 36/2 (1995), pp. 247-263.

Al-Ali, N., *Secularism, Gender and the State in the Middle East: the Egyptian's Women Movement*, Cambridge, Cambridge UP, 2000.

Alcott, L.M., *Little Women*, New York, Penguin, 1989.

Alessandri, A. d', "'La vie monastique dans l'Eglise orientale' di Dora D'Istria (1828-1888): problematiche religiose e politiche", *Trimestre*, 36/1-2 (2003), pp. 73-92.

Ali, A.H., "Strijd tegen absolute Allah is geboden", *NRC Handelsblad* (8 Nov. 2005).

Allan. T. & S. Gordon, *The Scalpel, the Sword: the Story of Doctor Norman Bethune*, New York, Monthly Review Press, 1973.

Amin, G., *'Asr al-tashhir bi-l-'arab wa-l-muslimin nahnu wa al-'alam ba'd 11 sabtambar 2001*, Al-Qahira, Dar al-Shuruq, 2004.

Andrade, J.C., *Obra poética completa*, Quito, Casa de la Cultura Ecuatoriana, 1976.

Anzulovic, B., *Heavenly Serbia: from Myth to Genocide*, London, Hurst, 1999.

Arsenidze, R., "Iz vospominanii o Staline", *The New Review. Novyi Zhurnal*, 72, June 1963, pp. 218-236.

Artizov, A. & O. Naumov (eds.), *Vlast' i Khudozhestvennaia Intelligentsiia. Dokumenty TsK RKP(b)-VKP(b), VChK-OGPU-NKVD o Kul'turnoi Politike. 1917-1953 gg.*, Moscow, Demokratiia, 1999.

Asmus, R., *Opening NATO's Door: how the Alliance Remade itself for a New Era*, New York, Columbia UP, 2002.

Atatürk, K., *Atatürk'ün Söylev ve Demeçleri II (1906-1938)*, Ankara, TTK, 1959.

Avedon, R., *Evidence, 1944-1994*, New York, Random House, 1994.

Bagrow, L., *History of Cartography*, London, Watts, 1964.

Bajon, J-Y., *Les années Mao – une histoire de la Chine en affiches (1949-1979)*, Paris, Les Éditions de Pacifique, 2001.

Baker, R., *Islam without Fear. Egypt and the New Islamists*, Cambridge Mass., Harvard UP, 2003.

Bakker, G., *Duitse geopolitiek 1919-1945: een imperialistiche ideologie*, Assen, 1967.

Baldick, C., *The Concise Oxford Dictionary of Literary Terms*, Oxford & New York, Oxford UP, 1990.

Balibar, E., "Y a-t-il un 'néo-racisme'?", in E. Balibar & I. Wallerstein, *Race, nation, classe. Les identités ambiguës*, Paris, Éditions La Découverte, 1988, pp. 27-41.

Barmé, G., "To Screw Foreigners is Patriotic: China's Avant-Garde Nationalists", *The China Journal*, 34 (July 1995).

Barnouw, D., *Oostboeren, zee-germanen en turfstekers*, Amsterdam, 2004.

Bayerlein, B.H. (ed.), *Georgii Dimitroff. Tagebücher 1933-1943*, Berlin, Aufbau, 2000.

Beardsell P.R., *Europe and Latin America: Returning the Gaze*, Manchester, Manchester UP, 2000.

Beardsell P.R., *Europe and Latin America: the Identity of the Other*, Manchester, University of Manchester, 1995.

Beardsell P.R., *Gods and Demons, Self and Other. Images of Europe in the Cultures of Latin America*, Hull, Hull UP, 1994.

Beller, M. & J. Leerssen (eds.), *Imagology*, Amsterdam, Rodopi, 2007.

Bevindingen over P. J. Meertens op grond van literatuur en geraadpleegde bronnen (Rapport van de Commissie van Drie), Amsterdam, 2006.

Bezymenskii, L.A., "Sekretnyi pakt s Gitlerom pisal lichno Stalin", *Novoe vremia*, 1, 1998, pp. 30-33.

Bhabha, H., *The Location of Culture*, London & New York, Routledge, 1994.

Biberaj, E., *Albania and China: a Study of an Unequal Alliance*, Boulder, Westview Press, 1986.

Bocsan, N. & I. Bolovan (eds.), *Calatori Români in Occident, Secolele XVII-XX*, Cluj-Napoca, Institutul Cultural Român, 2004.

Bohlen, C.E., *Witness to History. 1929-1969*, London, Weidenfeld & Nicolson, 1973.

Bojic, V., *Jacob Grimm und Vuk Karadzic: ein Vergleich ihrer Sprachauffassungen und ihre Zusammenarbeit auf dem Gebiet der serbischen Grammatik*, Munich, Otto Sagner, 1977.

Boldy, S., "Julio Cortázar, *Rayuela*", in P. Swanson (ed.), *Landmarks in Modern Latin American Fiction*, London and New York, Routledge, 1990.

Bolhuis, J. van, *Onderdrukking en verzet: Nederland in oorlogstijd*, Vol. I, Arnhem, 1949.

Boorstin, D.J., *America and the Image of Europe: Reflections on American Thought*, Gloucester Mass., Smith, 1976.

Borges, J.L., *Discusión*, Obras completas de Jorge Luis Borges 6, Buenos Aires, Emecé, 1961.

Bosman, K., *Ruimte voor een nieuwe tijd: vormgeving van de Nederlandse regio 1900-1945*, Rotterdam, 1993.

Bracewell, W. & A. Drace-Francis (eds.), *Balkan Travel Writing*, special issue, 6/1-2, 2005, of *Journeys: the International Journal of Travel and Travel Writing*.

Brady, A.-M., "Red and Expert: China's 'Foreign Friends' in the Great Proletarian Cultural Revolution, 1966-1969", in Woei Lien Chong (ed.), *China's Great Proletarian Cultural Revolution – Master Narratives and Post-Mao Counternarratives*, Lanham, Rowman & Littlefield, 2002, pp. 93-137.

Brandenberger, D., *National Bolshevism. Stalinist Mass Culture and the Formation of Modern Russian National Identity, 1931-1956*, Cambridge Mass., Harvard UP, 2002.

Breuer, S., *Anatomie der konservativen Revolution*, Darmstadt, Wissenschaftliche Buchgesellschaft, 1993.

Brock, P., *The Slovak National Awakening: an Essay in the Intellectual History of East Central Europe*, Toronto, Toronto UP, 1976.

Brotherston, G., *Image of the New World*, London, Thames and Hudson, 1979.

Buchwald, A., *I'll Always Have Paris! A Memoir*, New York, Fawcett Columbine, 1996.

Bugner, L. (ed.), *The Image of the Black in Western Art*, 2 Vols (in 3 parts), Cambridge Mass., Harvard UP, 1976-79.

Buncak, J. & M. Piscova, "Modern National Identity of Slovaks and their Attitude towards Europe", *Sociologia*, 32/3, 2000, pp. 289-310.

Buruma, I. & A. Margalit, *Occidentalism. The West in the Eyes of its Enemies*, New York, Penguin, 2004.

Calderón, E.C., *Americanos y europeos*, Madrid, Ediciones Guadarrama, 1957.

Carpenter, T.G., "The Domino Theory Reborn: Clinton's Bosnia Intervention and the 'Wider War' Thesis", *Cato Foreign Policy Briefing*, 42, 15 Aug. 1996, at http://www.cato.org/pubs/fpbriefs/fpb-042.html.

Chang-tai Hung, "The Fuming Image: Cartoons and Public Opinion in Late Republican China, 1945 to 1949", *Comparative Studies of Society and History*, 36/1, 1994.

Chen Tina Mai, "Female Icons, Feminist Iconography? Socialist Rhetoric and Women's Agency in 1950s China", *Gender & History*, 15/2, 2003, pp. 268-295.

Chen Xiaomei, *Occidentalism: a Theory of Counter-Discourse in Post-Mao China*, New York, Oxford UP, 1995.

Chen Zhimin, "Nationalism, Internationalism and Chinese Foreign Policy", *Journal of Contemporary China*, 14/42, Feb. 2005, pp. 36-37.

Chuev, F., *Sto Sorok Besed s Molotovym. Iz Dnevnika F. Chueva*, Moscow, TERRA, 1991.

Chuev, F., *Tak Govoril Kaganovich. Ispoved' Stalinskogo Apostola*, Moscow, Otechestvo, 1992.

Cisneros, A., *Como higuera en un campo de golf*, Lima, Instituto Nacional de Cultura, 1972.

Clark, K., *Petersburg. Crucible of Cultural Revolution*, Cambridge Mass., Harvard UP, 1995.

Coles, R., "James Baldwin Back Home", *New York Times Book Review*, 31 July 1977.

Cortázar, J., *Rayuela*, Buenos Aires & Mexico City, Alfaguara, 1966.

Crump, P. (ed.), *Atala, René*, Manchester, Manchester UP, 1961.

Cumhuriyet Halk Partisi, *On Beşinci Yıl Kitabı*, Ankara, 1938.

Danova, N., V. Dimova & M. Kalitsin (eds.), *Predstavata za 'Drugiya' na Balkanite*, Sofia, Marin Drinov, 1995.

Daskalov, R., *The Making of a Nation in the Balkans: Historiography of the Bulgarian Revival*, Budapest, Central European UP, 2004.

Deche, P. & C. Diamond, *Latin Americans in London: a Select List of Prominent Latin Americans in London, 1800-1996*, London, ILAS, University of London, 1998.

Delanty, G., *Inventing Europe: Idea, Identity, Reality*, Basingstoke, Macmillan, 1995.

Denton, K.A., "Visual Memory and the Construction of a Revolutionary Past: Paintings from the Museum of the Chinese Revolution", *Modern Chinese Literature and Culture*, 12/2, 2000.

Derks, H., *Deutsche Westforschung. Ideologie und Praxis im 20. Jahrhundert*, Leipzig, 2001.

Despatalovic, E.M., *Ljudevit Gaj and the Illyrian Movement*, Boulder, East European Quarterly, 1975.

Detrez, R., "Het Osmaanse *millet*-systeem", in R. Detrez & J. Blommaert (eds.), *Nationalisme: kritische opstellen*, Berchem, EPO, 1994, pp. 290-303.

DeWoskin, R., *Foreign Babes in Beijing – Behind the Scenes of a New China*, London, Granta, 2005.

Dietz, B. *et al.* (eds.), *Griff nach dem Westen. Die 'Westforschung' der völkisch-nationalen Wissenschaften zum nordwesteuropäischen Raum (1919-1960)*, 2 Vols, Munster, Waxmann, 2003.

Dikötter, F. (ed.), *The Construction of Racial Identities in China and Japan – Historical and Contemporary Perspectives*, London, Hurst, 1997.

Dikötter, F., *The Discourse of Race in Modern China*, London, Hurst, 1991.

Dilke, C. (ed.), *Letter to a King. A Picture-History of the Inca Civilisation by Huaman Poma*, London, George Allen & Unwin, 1978.

Ditt, K., "Franz Petri und die Geschichte der Niederlande. Vom germanischen Kulturraum zur Nation Europas", *Tijdschrift voor Geschiedenis*, 118/2, 2005, pp. 169-187.

Djilas, M., *Conversations with Stalin*, New York, Harcourt, Brace & World, 1962.

Droulia, L., *Philhellénisme. Ouvrages inspirés par la guerre de l'indépendance grecque, 1821-1833: Répertoire bibliographique*, Athens, Patris, 1974.

Dundy, E., *The Dud Avocado*, London, Virago, 2001 [1958].

Dupeux, L., *"Nationalbolschewismus" in Deutschland. 1919-1933: Kommunistische Strategie und konservative Dynamik*, Munich, Beck, 1976.

Edney, M.H., *The Origins and Development of J.B. Harley's Cartographic Theories*, Monograph 54, *Cartographica*, 40/1-2, 2005, Toronto, University of Toronto Press, 2005.

Elsie, R., *History of Albanian Literature*, 2 Vols, New York, Columbia UP, 1996.

Enepekides, P., "Beiträge zur geheimen geistigen und politischen Tätigkeit der Griechen in Wien vor der griechischen Revolution", *Berliner byzantinische Arbeiten*, 20, 1960.

Fallaci, O., *The Rage and the Pride*, New York, Rizolli, 2002.

Fanjul, S., "Islamofobia", *ABC*, 24 Feb. 2006.

Findley, C.V., *Bureaucratic Reform in the Ottoman Empire: the Sublime Porte, 1789-1922*, Princeton, Princeton UP, 1980.

Fleischhauer, I., "The Molotov-Ribbentrop Pact: the German Version", *International Affairs*, 8, 1991, pp. 114-129.

Foreign Relations of the United States. Diplomatic Papers. The Conferences at Malta and Yalta. 1945, Washington, US Government Printing Office, 1955.

Fraenkel-Verkade, E. (ed.), *Correspondentie Mr. M.M. Rost van Tonningen*, Vol. I, The Hague, 1967.

Frangou-Psychopedis, O., *I ethniki scholi mousikis: Provlimata ideoloyias*, Athens, Idryma Mesogeiakon Meleton, 1990.

Franolic, B., "The Development of Literary Croatian and Serbian", in I. Fodor & C. Hagège (eds.), *Language Reform: History and Future*, Hamburg, Buske, 1983-94, pp. 85-112.

Fredrickson, G.M., *The Comparative Imagination. On the History of Racism, Nationalism, and Social Movements*, Berkeley, University of California Press, 1997.

Fuentes, C., *El espejo enterrado*, Mexico, Fondo de Cultura Económica, 1992.

Furtado, C., *Economic Development of Latin America. Historical Background and Contemporary Problems*, Cambridge, Cambridge UP, 1976.

Galen Last, D. van, "Rassenhygiëne: de dodelijke omhelzing van wetenschap en nationaal-socialisme. Literatuuroverzicht", in *Volkseigen, ras, cultuur en wetenschap in Nederland 1900-1950; Elfde Jaarboek Oorlogsdocumentatie*, Zutphen, Walburg, 2000, pp. 213-247.

Geming da pipan baotou xuanji, Hangzhou, Zhejiang Renmin Meishu Chubanshe, 1970.

Gibianskii, L., "The Soviet Bloc and the Initial Stage of the Cold War: Archival Documents on Stalin's Meetings with Communist leaders of Yugoslavia and Bulgaria, 1946-1848", *Cold War International History Project Bulletin*, 10, 1998, pp. 112-134.

Gilbert, F., *To The Farewell Address: Ideas of Early American Foreign Policy*, Princeton, Princeton UP, 1961.

Gill, B. & M. Murphy, "China's Evolving Approach to Counterterrorism", *Harvard Asia Quarterly*, 9/1&2, Winter/Spring 2005.

Giunta, A., "Strategies of Modernity in Latin America", in G. Mosquera (ed.), *Beyond the Fantastic. Contemporary Art Criticism from Latin America*, London, Institute of International Visual Arts, 1995, pp. 53-66.

Glenny, M., "Only in the Balkans", *London Review of Books*, 29 May 1999.

Glenny, M., *The Balkans: Nationalism, War and the Great Powers 1804-1999*, London, Granta, 1999.

Göçek, F.M., *Rise of the Bourgeoisie, Demise of Empire: Ottoman Westernization and Social Change*, Oxford, Oxford UP, 1996.

Goldstein, L.J., "Return to Zhenbao Island: Who Started Shooting and Why it Matters", *The China Quarterly*, 168, Dec. 2001, pp. 985-997.

Goldsworthy, V., *Inventing Ruritania: the Imperialism of the Imagination*, New Haven, Yale UP, 1998.

Gommers, P.H., *Europe – What's in a Name?*, Leuven, Leuven UP, 2001.

Goodrick-Clarke, N., *Hitler's Priestess: Savitri Devi, the Hindu-Aryan Myth and Neo-Nazism*, New York, New York UP, 1998.

Gopnik, A., *Paris to the Moon: a Family in France*, London, Vintage, 2001.

Goss, J., *The Mapmaker's Art: a History of Cartography*, London, Studio Editions, 1993.

Greene, J.P., *The Intellectual Construction of America: Exceptionalism and Identity from 1492 to 1800*, Chapel Hill, University of North Carolina Press, 1993.

Gries, P.H., *China's New Nationalism – Pride, Politics, and Diplomacy*, Berkeley, University of California Press, 2004.

Gromyko, A.A., *Pamiatnoe*, Vol. 1, Moscow, Izdatel'stvo politicheskoi literatury, 1988.

Guerrina, R., *Europe: Histories, Ideas and Ideologies*, London, Arnold, 2002.

Günther, H., *Der sozialistische Übermensch. M. Gor'kij und der sowjetische Heldenmythos*, Stuttgart, Metzler, 1993.

Halfin, I., *Terror in my Soul. Communist Autobiographies on Trial*, Cambridge Mass., Harvard UP, 2003.

Hamm, J., "Randbemerkungen zur Entstehung und Entwicklung der serbokroatischen Schriftsprache", in C. Hannick (ed.), *Sprachen und Nationen im Bal-*

kanraum: die historischen Bedingungen der Entstehung der heutigen Nationalsprachen, Vienna, Böhlau, 1987, pp. 65-76.

Harbsmeier, M., "Character, Identity, and the Construction of Europe", *Ethnologia Europaea*, 29/2 (1999), pp. 5-12.

Harley, J.B. & D. Woodward (eds.), *The History of Cartography*, Vol. II, Book I, *Cartography in the Traditional Islamic and South Asian Societies*, Chicago, University of Chicago Press, 1992.

Harley, J.B. & D. Woodward (eds.), *The History of Cartography*, Vol. II, Book II, *Cartography in the Traditional East and Southeast Asian Societies*, Chicago, University of Chicago Press, 1994.

Harper, J.L., *American Visions of Europe: Franklin D. Roosevelt, George F. Kennan, and Dean G. Acheson*, New York, Cambridge UP, 1994.

Harvey, J.L., "Hans Rothfels: Issues and Paradoxes of an International Debate", *Sozial Geschichte*, 22/1 (2007), pp. 7-39.

Hemingway, E., *A Moveable Feast*, London, Arrow, 2004.

Herf, J., *Reactionary Modernism. Technology, Culture, and Politics in Weimar and the Third Reich*, Cambridge, Cambridge UP, 1984.

Hilal, R., *Al-Amraka wa al-Aslama: Ma'zaq 'Arab al-Yaum*, Al-Qahira, Dar Masr al-Mahrusa, 2004.

Hilger, G., *Wir und der Kreml. Deutsch-sowjetische Beziehungen 1918-1941. Erinnerungen eines deutschen Diplomaten*, Frankfurt, Alfred Metzner, 1955.

Hillenbrand, C., *The Crusades: Islamic Perspectives*, Edinburgh, Edinburgh UP, 1999.

Hirsch, F., *Empire of Nations. Ethnographic Knowledge and the Making of the Soviet Union*, Ithaca, Cornell UP, 2005.

Hirsch, F., "Race without the Practice of Racial Politics", *Slavic Review*, 61/1, 2002, pp. 30-43.

Hirshler, E.E. "At Home in Paris" in K. Adler, E.E. Hirshler, H.B. Weinberg, *Americans in Paris, 1860-1900*, London, National Gallery, 2006.

Hitchens, C., *Blood, Class and Empire: the Enduring Anglo-American Relationship*, New York, Nation Books, 2004.

Hollander, A.N.J. den, "As Others see Us", *Synthese*, 6/5&6, May 1947-May 1948, pp. 214-237.

Hourani, A., *Arabic Thought in the Liberal Age, 1798-1939*, Cambridge, Cambridge UP, 1983.

Hoxha, E., *With Stalin. Memoirs*, Tirana, 8 Nëntori, 1979.

Ibor, J.J.L., *El Español y su complejo de inferioridad*, Madrid, Rialp, 1951.

Imara, M., *Al-islam wa al-gharb. Ayna al-khat'a? .. wa ayna al-sawab?*, Al-Qahira, Maktaba al-shuruq al-dawliyya, 2004.

Islami, M., *Naum P. Veqilharxhi*, Prishtinë, Rilindja, 1969.

"I.V. Stalin v rabote nad 'Kratkim kursom istorii VKP(b)'. Okonchanie", *Voprosy istorii*, 4, 2003, pp. 3-25.

James, H., *The American Scene*, London, Chapman & Hall, 1907.

James, H., *The Portrait of a Lady*, Oxford, Oxford UP, 1995 [1881].

Jelavich, B., *History of the Balkans: Eighteenth and Nineteenth Centuries*, Cambridge, Cambridge UP, 1983.

Jelavich, C. & B. (eds.), *The Balkans in Transition*, Berkeley, University of California Press, 1963.

Johansson, P., "White Skin, Large Breasts: Chinese Beauty Product Advertising as Cultural Discourse", *China Information*, 13/2&3 (Autumn/Winter 1998).

Jones, J., *A Common Place. The Representation of Paris in Spanish American Fiction*, Lewisburg, Bucknell UP, 1998.

Kagan, R., *Paradise and Power: America and Europe in the New World Order*, London, Atlantic, 2003.

Kaplan, R.D., *Balkan Ghosts: a Journey through History*, New York, St Martin's Press, 1993.

Keijzer, J., "Haushofer in Holland. Een studie naar de geopolitieke visie binnen de NSB", unpublished *doctoraalscriptie* [MA thesis] Political Geography, University of Amsterdam, 2002.

Kennedy-Pipe, C., *Stalin's Cold War. Soviet Strategies in Europe, 1943 to 1956*, Manchester, Manchester UP, 1995.

Kershaw, I., *Hitler 1936-1945. Nemesis*, London, 2000.

Kissinger, H., *Years of Upheaval*, Boston, Little Brown, 1982.

Kline, G.L., *Religious and Anti-Religious Thought in Russia. The Weil Lectures*, Chicago, University of Chicago Press, 1968.

Knock, T.J., *To End All Wars: Woodrow Wilson and the Quest for a New World Order*, Princeton, Princeton UP, 1992.

Kohn, H., *Pan-Slavism, its History and Ideology*, New York, Vintage, 1960.

Kolev, N., "La collaboration entre le Dr Nicolas S. Piccolos et Claude Fauriel", *Etudes Balkaniques*, 2, 1978, pp. 114-123.

Konstantinou, E. (ed.), *Europäischer Philhellenismus: die europäische philhellenische Literatur bis zur 1. Hälfte des 19. Jahrhunderts*, Frankfurt, Peter Lang, 1992.

Konstantinovic, Z., "'Tirk oder Griech'. Zur Kontamination ihrer Epitheta", in F.K. Stanzel *et al.* (eds.), *Europäischer Völkerspiegel: imagologisch-ethnographische Studien zu den Völkertafeln des frühen 18. Jahrhunderts*, Heidelberg, Carl Winter, 1999, pp. 299-314.

Kostova, L., *Tales of the Periphery: the Balkans in Nineteenth-Century British Writing*, Veliko Turnovo, Veliko Turnove UP, 1997.

Krementsov, N., *Stalinist Science*, Princeton, Princeton UP, 1997.

Laing, E.J., *Selling Happiness – Calendar Posters and Visual Culture in Early Twentieth-Century Shanghai*, Honolulu, University of Hawai'i Press, 2004.

Landsberger, S., "Learning by what Example? Educational Propaganda in Twenty-First-Century China", *Critical Asian Studies*, 33/1, 2001.

Landsberger, S., *Paint it Red – Fifty Years of Chinese Propaganda Posters*, Groningen, Intermed, 1998.

Lawson, T., "The Myth of the European Civil War", in R. Littlejohns & S. Soncini (eds.), *Myths of Europe*, Amsterdam, Rodopi, 2007, pp. 275-289.

Leader of People – Collection of Works at the National Art Exhibition in Celebration of the 70th Birthday of Great Leader President Kim Il Sung, Pyongyang, Foreign Languages Publishing House, 1984.

Lee Chin-Chuan, Zhou He & Yu Huang, "'Chinese Party Publicity Inc.' Conglomerated: the Case of the Shenzhen Press Group", *Media, Culture & Society*, 28/4, 2006.

Leerssen, J.T., "Een omzwerving rond Monte-Cristo. Mediterrane beeldvorming van Byron tot Bogart en Braudel", in A.M. van Erp Taalman Kip & I.F. de Jong (eds.), *Schurken en schelmen: cultuurhistorische verkenningen rond de Middellandse Zee*, Amsterdam, Amsterdam UP, 1995, pp. 9-24.

Leithäuser, J.G., *Mappae Mundi: die geistige Eroberung der Welt*, Berlin, Safari [1958].

León-Portilla, M. (ed.), *The Broken Spears. The Aztec Account of the Conquest of Mexico*, Boston, Beacon Press, 1992.

Lewis, B., *From Babel to Dragomans: Interpreting the Middle East*, Oxford, Oxford UP, 2004.

Lewis, B., *What Went Wrong? Western Impact and Middle Eastern Response*, Oxford, Oxford UP, 2002.

Li Jingying, "Huang Jiguang", *Qingnian yingxiongde gushi*, Beijing, Zhongguo Qingnian Chubanshe, 1965.

Li Ming, *Huang Jiguang*, Beijing, Zhongguo Shaonian Ertong Chubanshe, 1957.

Lindheim, R. & G.S.N. Luckyj (eds.), *Towards an Intellectual History of the Ukraine: an Anthology of Ukrainian Thought from 1710 to 1995*, Toronto, Toronto UP, 1996.

Littlejohns, R. & S. Soncini (eds.), *Myths of Europe*, Amsterdam, Rodopi, 2007.

Loohuis, J.G., *Landbouwgordel en de industriecentra der gematigde zone in hare beteekenis voor het huidige aspect van het Koloniale vraagstuk in Zuid-Oost-Azië*, Hilversum, 1931.

Lord, A.B., "Nationalism and the Muses in the Balkan Slavic Literature in the Modern Period", in C. & B. Jelavich (eds.), *The Balkans in Transition*, Berkeley, University of California Press, 1963, pp. 258-296.

Lorenz, C., "De Westforschung – een geval apart?", *Tijdschrift voor Geschiedenis*, 118/2, 2005, pp. 252-271.

Luna, F., *Breve historia de los argentinos*, 14th edition, Buenos Aires, Planeta, 1997.

Mackerras, C., *Western Images of China*, Hong Kong, Oxford UP, 1989.

MacMillan, M., *Paris, 1919: Six Months that Changed the World*, New York, Random House, 2001.

Maksimenkov, L., "Slavianskie natsii – molodye i zdorovye", *Nezavisimaia Gazeta*, 31 Jan. 2006.

Malyshev, V.A., "Dnevnik narkoma. 'Proidet desiatok let, i eti vstrechi ne vosstanovish' uzhe v pamiati'", *Istochnik*, 5, 1997, pp. 103-147.

Mao Tse-tung, *Selected Works of Mao Tse-tung*, Vol. II, Peking, Foreign Languages Press, 1965.

Mar'ina, V.V., "Peregovory E. Benesha v Moskve (dekabr' 1943 g.)", *Voprosy Istorii*, 3, 2001, pp. 3-20.

Mardin, Ş., *The Genesis of Young Ottoman Thought. A Study in the Modernization of Turkish Political Ideas*, Princeton, Princeton UP, 1962.

Margaritis, G., "Griechenland: Wiedergeburt aus dem Geist der Antike", in M. Flacke (ed.), *Mythen der Nationen: ein europäisches Panorama*, Munich, Koehler & Amelang, 1998, pp. 152-173.

Márquez Rodríguez, A., "Deslinde entre el *realismo mágico* y *lo real maravilloso* a propósito de la novelística de García Márquez", in A. María Hernández de López (ed.), *En la punta de la Mira: Gabriel García Márquez*, Madrid, Editorial Pliegos, 1985.

Márquez, G.G., "La soledad de la América Latina", *Casa de las Américas*, 23, No. 137, 1983.

Martin, T., *The Affirmative Action Empire. Nations and Nationalities in the Soviet Union, 1923-1939*, Ithaca, Cornell UP.

McCune, S., "Korean Maps of the Yi Dynasty", *Korean Culture*, 4/3, Sept. 1983, pp. 21-31.

McGowan, M., "Some Turkish-German Views and Visions of 'Europe'", in B. Axford *et al.* (eds.), *Unity and Diversity in the New Europe*, Oxford, Peter Lang, 2000, pp. 339-369.

Mehnert, K., *Stalin versus Marx. The Stalinist Historical Doctrine*, London, George Allen & Unwin, 1952.

Merchiers, I., "Cultural Nationalism in the South Slav Habsburg Lands in the Early Nineteenth Century: the Scholarly Network of Jernej Kopitar (1780-1844)", doctoral thesis, Universiteit Gent, Faculteit Letteren en Wijsbegeerte, 2005.

Milisavac, Z., *Matica Srpska, 1829-1964*, Novi Sad, Matica Srpska, 1965.

Miller, H., *Tropic of Cancer*, Paris, Obelisk, 1957.

Milojkovic-Djuric, J., *Panslavism and National Identity in Russia and in the Balkans, 1830-1880: Images of the Self and Others*, New York, Columbia UP, 1994.

Mohler, A., *Die Konservative Revolution in Deutschland. 1918-1932*, Stuttgart, Friedrich Vorwerk, 1950.

Moreh, S., *Al-Jabarti's Chronicle of the First Seven Months of the French Occupation of Egypt*, Leiden, Brill, 1975.

Morley, D. & K. Robins, *Spaces of Identity – Global Media, Electronic Landscapes and Cultural Boundaries*, London, Routledge, 1995.

Morris, A., "'I Believe You can Fly': Basketball Culture in Postsocialist China", in P. Link, R.P. Madsen & P.G. Pickowicz (eds.), *Popular China – Unofficial Culture in a Globalizing Society*, Lanham, Rowman & Littlefield, 2002.

Moser, D., "Media 'Schizophrenia' in China", www.danwei.org (accessed 7 Aug. 2006).

Müller, R.-D. (ed.), *Die deutsche Wirtschaftspolitik in den besetzten sowjetischen Gebieten 1941-1943. Der Abschlussbericht des Wirtschaftstabes Ost und Aufzeichnungen eines Angehörigen des Wirtschaftskommandos Kiew*, Boppard am Rhein, 1991.

Murin, I.G., "'Nevol'niki v rukakh germanskogo Reikhsvera': Rech' I.V. Stalina v Narkomate oborony", *Istochnik*, 3, 1994, pp. 72-88.

Nafi', H., *Al-Ittihad al-Urubbi wa al-durus al-mustafada 'arabiyyan*, Bayrut, Markaz dirasat al-wahda al-'arabiyya, 2004.

Nafi, I. (ed.), *Madha yagri fi Sharq Urubba?*, Al-Qahira, Al-Ahram 2001.

Nevezhin, B.A., *Zastol'nye Rechi Stalina. Dokumenty i Materialy*, Moscow, AIRO-XX, 2003.

Newman, D.L., *An Imam in Paris: Al-Tahtawi's visit to France (1826-31)*, London, Saqi Books, 2004.

Ngai, M.M., *Impossible Subjects: Illegal Aliens and the Making of Modern America*, Princeton, Princeton UP, 2005.

Niscov, V., "Über den Widerhall der volkskundlichen Beschäftigung der Brüder Grimm in Rumänien", *Brüder Grimm Gedenken*, 2, 1975, pp. 146-167.

Noe, A. (ed.), *Der Philhellenismus in der westeuropäischen Literatur 1780-1830*, Amsterdam, Rodopi, 1994.

Nurmi, I., *Slovakia: a Playground for Nationalism and National Identity: Manifestations of the National Identity of Slovaks, 1918-1920*, Helsinki, Suomen Historiallinen Seura, 1999.

"'Nuzhno idti k sotsializmuy ne priamo, a zigzagami'. Zapis' besedy I.V. Stalina s rukovoditeliami SEPG. Dekabr' 1948 g.", *Istoricheskii arkhiv*, 5, 2002, pp. 3-26.

Nyíri, P., "The Yellow Man's Burden: Chinese Migrants on a Civilizing Mission", *The China Journal*, 56, July 2006.

Omissi, D.E., "Europe through Indian Eyes", *English Historical Review*, 122/2, 2007, pp. 371-96.

Passerini, L., *Figures d'Europe: Images and Myths of Europe*, Brussels, PIE Peter Lang, 2003.

Path of Great Love, The, Pyongyang, Foreign Languages Publishing House, 1977.

Pearton, M., "Nicolae Iorga as Historian and Politician", in D. Deletant & H. Hanak (eds.), *Historians as Nation-Builder: Central and South-East Europe*, London, Macmillan/SSEES, 1988, pp. 157-173.

Pechenkin, A.A., "'Sovremennaia armiia – armiia nastupatel'naia'. Vystupleniia I.V. Stalina na prieme v Kremle pered vypusnikami voennykh akademii. Mai 1941 g.", *Istoricheskii arkhiv*, 2, 1995, pp. 23-31.

Pells, R., *Not Like Us: How Europeans have Loved, Hated, and Transformed American Culture since World War II*, New York, Basic Books, 1997.

Petri, F., *Die fränkische Landnahme und die Entstehung der germanisch-romanischen Sprachgrenze in der interdisziplinären Diskussion*, Darmstadt, 1977.

Petri, F., *Germanisches Volkserbe in Wallonien und Nordfrankreich: die fränkische Landnahme in Frankreich und den Niederlanden und die Bildung der westlichen Sprachgrenze*, 2 Vols, Bonn, 1937 & 1942.

Petri, F., *Die Niederlände (Holland und Belgien) und das Reich. Volkstum, Geschichte und Gegenwart*, Bonn, 1940.

Pipes, D., "The Western Mind of Radical Islam", *First Things*, 58, 1995, pp. 18-23.

Pocock, J.G.A., *The Machiavellian Moment: Florentine Political Thought and the Atlantic Republican Tradition*, Princeton, Princeton UP, 1975.

Prel, M.F. du & W. Janke (eds.), *Die Niederlande in Umbruch der Zeiten: Alte und Neue Beziehungen zum Reich*, Würzburg, 1941.

Priestland, D., "Stalin as Bolshevik Romantic: Ideology and Mobilisation, 1917-1919", in S. Davies & J. Harris (eds.), *Stalin. A New History*, Cambridge, Cambridge UP, 2005, pp. 181-201.

Quijano, A., "Modernity, Identity, and Utopia in Latin America", *Boundary 2*, 20/3, 1993, pp. 140-155.

Quiroga, H., *El salvaje y otros cuentos*, 3rd edition, Buenos Aires, Losada, 1976.

Rama, A., *Transculturación narrativa en América Latina*, Montevideo, Arca, 1989.

Reed, C.A., "Re/Collecting the Sources: Shanghai's *Dianshizhai Pictorial* and Its Place in Historical Memories, 1884-1949", *Modern Chinese Literature and Culture*, 12/2, 2000.

Reiter, N. (ed.), *Nationalbewegungen auf dem Balkan*, Wiesbaden, Harassowitz, 1983.

Richard, C., *The Founders and the Classics: Greece, Rome, and the American Enlightenment*, Cambridge MA, Harvard UP, 1995.

Rogaski, R., "Nature, Annihilation, and Modernity: China's Korean War Germ-Warfare Experience Reconsidered", *Journal of Asian Studies*, 61/2, 2002, pp. 381-415.

Rose, P.I., *They and We: Racial and Ethnic Relations in the United States*, 6th edition, Boulder, Paradigm, 2006.

Rosenthal, B. (ed.), *New Myth, New World. From Nietzsche to Stalinism*, University Park Pa., Pennsylvania State UP, 2002.

Rosenthal, B. (ed.), *Nietzsche in Russia*, Princeton, Princeton UP, 1986.

Rössler, M. & S. Schleiermacher (eds.), *Der "Generalplan Ost" Hauptlinien der nationalsozialistischen Planungs- und Vernichtungspolitik*, Berlin, 1993.

Rowe, W. & V. Schelling, *Memory and Modernity. Popular Culture in Latin America*, London & New York, Verso, 1991.

Saich, T., *The Origins of the First United Front in China – The Role of Sneevliet (Alias Maring)*, 2 Vols, Leiden, Brill, 1991.

Sampimon, J., *Becoming Bulgarian: the Articulation of Bulgarian Identity in the Nineteenth Century in its International Context, an Intellectual History*, Amsterdam, Pegasus, 2006.

Sarmiento, D.S., *Facundo o la civilización y la barbarie*, 1845.

Sautman, B., "Anti-Black Racism in Post-Mao China", *The China Quarterly*, 138, June 1994, pp. 429-434.

Schaeffer, W., "Shanghai Savage", *Positions*, 11/1, 2003, pp. 91-133.

Schmaltz, E. & S. Sinner, "From Genealogy to Genocide and back to Genealogy", in I. Haar & M. Falbusch (eds.), *German Scholars and Ethnic Cleansing, 1920-1945*, New York, 2005, pp. 51-85.

Schoenhals, M., *Doing Things with Words in Chinese Politics*, Berkeley, Institute of East Asian Studies, 1992.

Schöffer, I., *Het nationaal-socialistische beeld van de geschiedenis der Nederlanden*, Arnhem/Amsterdam [1956].

Schoonbaert, L. *et al.* (eds.), *America Bride of the Sun: 500 Years Latin America and the Low Countries*, Ghent, Imschoot, 1992.

Schroeder, K.-H., "Marginale Romania: die Bildung und Normierung der rumänischen Schriftsprache", in C. Hannick (ed.), *Sprachen und Nationen im Balkanraum: die historischen Bedingungen der Entstehung der heutigen Nationalsprachen*, Vienna, Böhlau, 1987, pp. 195-209.

Schüddekopf, O., *Linke Leute von rechts. Die nationalrevolutionäre Minderheiten und der Kommunismus in der Weimarer Republik*, Stuttgart, Kohlhammer, 1960.

Schulze, W. & O.G. Oexle (eds.), *Deutsche Historiker im Nationalsozialismus*, Frankfurt am Main, 1999.

Shahin, E.E., *Through Muslim Eyes: M. Rashid Rida and the West*, Herndon Va, International Institute of Islamic Thought, 1993.

Shao Wu *et al.* (eds.), *Gongheguo qunyingpu*, Beijing, Zhongguo Shaonian Ertong Chubanshe, 2003.

Shaw, S., *Between Old and New. The Ottoman Empire under Sultan Selim III 1789-1807*, Cambridge Mass., Harvard UP, 1971.

Shen Qian & Zhong Shi, *Shiyong meishu zi hua*, Shanghai, Tongren Shushi, 1951.

Shen Tongheng, *Zenyang xue manhua*, Shanghai, Silian Chubanshe, 1955.

Shepilov, D.T., "Vospominaniia", *Voprosy istorii*, 5, 1998, pp. 3-27.

Shipley, J.T. (ed.), *Dictionary of World Literary Terms*, London, George Allen & Unwin, 1970.

Short, D., "The Use and Abuse of the Language Argument in mid-Nineteenth Century 'Czechoslovakism': an Appraisal of a Propaganda Milestone", in R.B. Pynsent (ed.), *The Literature of Nationalism: Essays on East European Identity*, London, Macmillan/SSEES, 1996, pp. 40-65.

Sieferle, R.P., *Die Konservative Revolution. Fünf biografische Skizzen*, Frankfurt M., Fischer Taschenbuch, 1995.

Sipols, V.I., "Missiia Krippsa v 1940 g. Beseda so Stalinym", *Novaia i noveishaia istoriia*, 5, 1992, pp. 23-40.

Siqueiros, D.A., *Art and Revolution*, translated by S. Calles, London, Lawrence & Wishart, 1975.

Slezkine, Y., *Arctic Mirrors. Russia and the Small Peoples of the North*, Ithaca, Cornell UP, 1994.

Snow, P., *The Star Raft – China's Encounter with Africa*, London, Weidenfeld & Nicolson, 1988.

Sombart, W., *Händler und Helden. Patriotische Besinnungen*, Munich/Leipzig, Duncker & Humblot, 1915.

Sontag, R.J. & J.S. Beddie (eds.), *Nazi-Soviet Relations. 1939-1941. Documents from the Archives of the German Foreign Office as Released by the Department of State*, New York, Didier, 1948.

Spencer, T., *Fair Greece, Sad Relic: Literary Philhellenism from Shakespeare to Byron*, Athens, Harvey, 1986.

Stalin, I.V., *Sochineniia*, Vols 1-13, Moscow, Gosudarstvennoe izdatel'stvo politicheskoi literatury, 1946-1951; Vols 1[XIV]-3[XVI], Stanford, Hoover Institution on War, Revolution, and Peace, 1967.

"Stalin's Conversations with Chinese leaders. Talks with Mao Zedong. December 1949-January 1950, and with Zhou Enlai, August-September 1952", *Cold War International History Project Bulletin*, 6-7, 1995/96, pp. 4-29.

Stanzel, F.K. *et al.* (eds.), *Europäischer Völkerspiegel: imagologisch-ethnographische Studien zu den Völkertafeln des frühen 18. Jahrhunderts*, Heidelberg, Carl Winter, 1999.

Stoianovich, T., *Balkan Worlds: the First and Last Europe*, Armonk, NY, Sharpe, 1994.

Stoianovich, T., *Between East and West: the Balkan and Mediterranean Worlds*, New Rochelle NY, Caratzas, 1992.

Stråth, B. (ed.), *Europe and the Other and Europe as the Other*, Brussels, PIE Peter Lang, 2000.

Sullivan, M.J., "The 1988-89 Nanjing Anti-African Protests: Racial Nationalism or National Racism?", *The China Quarterly*, 138, June 1994.

Sýkora, V., "Vooruit, kinderen der Grieken...: Politisering en mythologisering van de Griekse nationale held Rigas Velestinlis Feraios", MA thesis, Universiteit Utrecht, History, 2005.

Tames, I., '*Oorlog voor onze gedachten': oorlog, neutraliteit en identiteit in het Nederlandse publieke debat, 1914-1918*, Hilversum, 2006.

Tanner, M., *Croatia: a Nation Forged in War*, New Haven, Yale UP, 1997.

Testen, P., "Žiga Zois: Biografska skica", *Glasnik Slovenske Matice*, 29-31, 2005-07, pp. 118-125.

Theunisz, J., *Bibliographie van Belgische en Nederlandsche geschriften betreffende vestiging van Nederlanders in Midden- en Oost-Europa*, Amsterdam, 1941.

Theunisz, J., *Naar Oostland. Nederlandse kolonisatie in Europa*, Amsterdam, 1943.

Thomas, J. (ed.), *Reading Images*, London, Palgrave, 2000.

Todorova, M., *Imagining the Balkans*, Oxford, Oxford UP, 1997.

Troyen, C., "Innocents Abroad: American Painters at the 1867 Exposition Universelle, Paris", *American Art Journal*, 16/4, 1984, pp. 2-29.

Truman, H.S., *Memoirs: Years of Trial and Hope, 1946-1952*, New York, Signet Books, 1956.

Tuchman, B.W., *Stilwell and the American Experience in China 1911-1945*, New York, Macmillan, 1971.

Tucker, R., *The Soviet Political Mind. Stalinism and Post-Stalin Change*, revised ed., London, George Allen & Unwin, 1972.

Twain, M., *Innocents Abroad & Roughing It*, New York, Library of America, 1984.

"'Videl, kak voznikaet kul't lichnosti. Mat'iash Rakoshi o Staline i o sebe", *Istochnik*, 1, 1997, pp. 111-140.

Volkogonov, D., *Sem' vozhdei. Galeria liderov SSSR v 2-kh knigakh*, Moscow, Novosti, 1995.

Volkov, S., *Shostakovich and Stalin. The Extraordinary Relationship between the Great Composer and the Brutal Dictator*, London, Little, Brown, 2004.

Volkseigen, ras, cultuur en wetenschap in Nederland 1900-1950; Elfde Jaarboek Oorlogsdocumentatie, Zutphen, 2000.

Volokitina, T.V. *et al.*, *Sovetskii Faktor v Vostochnoi Evrope. 1944-1953*, Vol. 1, *1944-1948*, Moscow, ROSSPEN, 1999.

Volokitina, T.V. *et al.*, *Vostochnaia Evropa v Dokumentakh Rossiiskikh Arkhivov. 1944-1953 gg.*, Vol. 1, *1944-1948 gg.*, Novosibirsk, Sibirskii khronograf, 1997.

Voorrecht, R., 'Westerlingen in China – een onderzoek naar de berichtgeving in de Dianshizhai huabao 1884-1886, 1896-1898', unpublished MA thesis, Leiden University, 1987.

Wachtel, A.B., *Making a Nation, Breaking a Nation: Literature and Cultural Politics in Yugoslavia*, Stanford, Stanford UP, 1998.

Wachtel, N., *La Vision des vaincus. Les Indiens du Pérou devant la conquête espagnole 1530-1570*, Paris, Gallimard, 1971.

Wang Xiaohua & Sun Qing, *Bainian shenghuo bianqian*, Nanjing, Jiangsu Meishu Chubanshe, 2000.

Wauchope, R. (ed.), *Handbook of Middle American Indians*, Austin, University of Texas Press, 1964-75.

Wegelin, C., *The Image of Europe in Henry James*, Dallas, Southern Methodist UP, 1958.

Weinberg, H.B., "Summers in the Country", in K. Adler, E.E. Hirshler, H.B. Weinberg, *Americans in Paris, 1860-1900*, London, National Gallery, 2006.

Weiner, A., "Nothing but Certainty", *Slavic Review*, 61/1, 2002, pp. 44-53.

Weitz, E.D., "Racial Politics without the Concept of Race: Reevaluating Soviet Ethnic and National Purges", *Slavic Review*, 61/1, 2002, pp. 1-29.

Weststeijn, W.G., *Russische literatuur*, Amsterdam, Meulenhoff, 2004.

Wharton, E., *A Backward Glance: an Autobiography*, New York, Touchstone, 1998.

Whitfield, P., *The Image of the World: 20 Centuries of World Maps*, London, British Library, 1994.

Wilson, D., *The Life and Times of Vuk Stefanovic Karadzic, 1787-1864: Literacy, Literature and National Independence in Serbia*, Oxford, Clarendon Press, 1970.

Wintle, M.J. (ed.), *Culture and Identity in Europe*, Aldershot, Avebury Press, 1996.

Wintle, M.J., *The Image of Europe: Visualizing Europe in Cartography and Iconography*, Cambridge, Cambridge UP, 2008 forthcoming.

Wolff, L., *Inventing Eastern Europe: the Map of Civilization on the Mind of the Enlightenment*, Stanford, Stanford UP, 1994.

Wolff, L., *Venice and the Slavs: the Discovery of Dalmatia in the Age of Enlightenment*, Stanford, Stanford UP, 2001.

Woltering, R., "Ik dacht dat de UNDP het nooit zou publiceren", *ZemZem*, 1/2, 2005.

Wood, G.S., *The Radicalism of the American Revolution*, New York, Vintage, 1993.

Woods, R., *The Conservative Revolution in the Weimar Republic*, Basingstoke, Palgrave, 2001.

Xu Yan, "Yang Gensi", *80 wei gongchandang rende gushi*, Beijing, Jiefangjun Wenyi Chubanshe, 2001.

Xuanchuanhua cankao ziliao huibian, Beijing, Renmin Meishu Chubanshe, 1953.

Ye, Xiaoqing, *The Dianshizhai Pictorial – Shanghai Urban Life 1884-1898*, Ann Arbor, University of Michigan Center for Chinese Studies, 2003.

Zabel, M.D. (ed.), *The Portable Henry James*, New York, Viking, 1960.

Zaitsev, P., *Taras Shevchenko: a Life*, transl. by G.S.N. Luckyj, Toronto, Toronto UP, 1988.

"Zapis' besedy I.V. Stalina i Sharlia de Gollia", *Istochnik*, 5, 1996.

"Zapis' besedy tov I.V. Stalina s sekretarem TsK Frantsuzskoi kompartii Torezom", *Istoricheskii arkhiv*, 1, 1996, pp. 6-26.

"Zapis' besedy tov. I.V. Stalina s Cherchillem. 9 oktiabria 1944 g. v 22 chasa", *Istochnik*, 2, 2003, pp. 48-53.

Zürcher, E-J., "Two Young Ottomanists Discover Kemalist Turkey. The Travel Diaries of Robert Anhegger and Andreas Tietze", *Journal of Turkish Studies*, 26/1, 2002, pp. 359-369.

Zürcher, E-J., "Young Turks, Ottoman Muslims and Turkish Nationalists: Identity Politics 1908-1938", in K.H. Karpat (ed.), *Ottoman Past and Today's Turkey*, Leiden, Brill, 2000, pp. 150-179.

Index

M

N

O

P

Q

R

S

Z

Brief Details on the Contributors

David Barnouw studied Political Science at the University of Amsterdam. Since 1979 he has been a researcher, staff member and Press Information Officer at the Netherlands Institute for War Documentation. This Institute deals with World War II, the German Occupation of the Netherlands and the Japanese Occupation of the Dutch East Indies. He has written books, articles and reviews about the Occupation period, and is specially interested in the Nazi Youth Movement in the Netherlands, and collaboration problems. He has also written about movies – both fiction and non-fiction – dealing with the Occupation. He is frequently invited to speak on national and international radio and television as an expert on the Second World War. With two colleagues he published *The Diary of Anne Frank: The Critical Edition* (1986), to great acclaim: 20,000 copies were sold within a year. This edition has been translated into German, English, French, Japanese and Italian.

Emeritus Professor **Peter Beardsell** graduated in the University of Manchester in French and Spanish, and wrote masters and doctoral theses on Latin American literature. He held lectureships in the UK Universities of Manchester and Sheffield, and the chair of Hispanic Studies in the University of Hull. In the national context he was President of the Society for Latin American Studies, and a member of the Iberian and Latin American Languages panels of the Research Assessment Exercises in 1996 and 2001. Publications include books on Latin American novelists, poets and dramatists, and the monograph *Europe and Latin America: Returning the Gaze* (Manchester University Press, 2000). His present research is focussed on Mexico's Teatro del Norte, with special reference to history, society, and the uncertainty principle.

Ruud Janssens is Professor and chair of American Studies at the University of Amsterdam. He taught at the Royal Netherlands Naval College, was an Advanced Research Fellow at the Weatherhead Center for International Affairs at Harvard University, and the 2007 Erasmus Lecturer there as well. Among his publications are *Post-Cold War Europe, Post-Cold War America* (co-edited with Rob Kroes; Amsterdam, 2004), *Of Mice and Men: American Imperialism and American Studies* (Amsterdam, 2004), and *'What Future for Japan?' U.S. Wartime Planning for the Postwar Era, 1942-1945* (Amsterdam, 1995).

Stefan R. Landsberger took his PhD in Sinology from Leiden University, the Netherlands. He holds the Olfert Dapper Chair of Contemporary Chinese Culture at the University of Amsterdam and is an associate professor at the Sinological Institute, Leiden University. He has published widely in books and journals, including *Critical Asian Studies*, *China Journal*, *China Review*, *China Information*, *Pacific Review* and others. Landsberger has one of the largest private collections of Chinese propaganda posters in the world. He maintains an extensive website exclusively devoted to this genre of political communications (http://www.iisg.nl/~landsberger). His research interests include various media-related aspects of the emerging Chinese urban consumer culture.

Joep Leerssen studied in Aachen, Dublin and Toronto, and now holds the Chair of Modern European Literature at the University of Amsterdam; he is also director of the Huizinga Institute (the Dutch national research institute for cultural history), and served as the 2003 Erasmus Lecturer at Harvard University. He has published widely on national identity, national stereotypes and cultural nationalism; following his work in Irish and Dutch-Flemish national thought he has conducted a research group (sponsored by the Dutch national funding agency NWO) on early nineteenth-century cultural nationalism in the Balkans. These projects form part of an ongoing programme on the network-carried spread of romantic nationalism in Europe.

Erik van Ree studied sociology at the University of Amsterdam. His PhD thesis, *Socialism in One Zone. Stalin's Policy in Korea, 1945-47*, was published in 1989 by Berg Publishers, Oxford. He is at present assistant professor in the department of European Studies at the University of Amsterdam. His other publications include: *The Political Thought of Joseph Stalin. A Study in Twentieth-Century Revolutionary Patriotism* (Routledge Curzon, 2002). His current research interest focuses on Stalin's revolutionary career before 1917.

Michael Wintle studied at Cambridge and Ghent Universities, and is Professor of European History at the University of Amsterdam, where he directs the degree programmes in European Studies. Prior to 2002, he held a chair of European History at the University of Hull, UK, where he had taught since 1980. His current research interests are in European identity and especially the visual representation of Europe, cultural aspects of European integration, European industrialisation, and the modern social and economic history of the Low Countries.

Robbert A.F.L. Woltering studied Arabic and Political Science at the University of Amsterdam, and is currently a PhD fellow at the International Institute for the Study of Islam in the Modern World (ISIM), in Leiden. His research concerns images of the West in contemporary Egyptian cultural discourse. He is an editor of *ZemZem*, the

Dutch periodical on the Middle East, North Africa and Islam. He recently published *Architectuur in Egypte* (with Jorrit Kelder), on his hobby, Egyptian architecture.

Erik-Jan Zürcher studied Turkish (with Arabic, Persian and Modern History) at the University of Leiden, where he also took his PhD in 1984. His thesis, *The Unionist Factor*, concerned the role of the Young Turk Committee of Union and Progress in the Turkish independence struggle after World War I. In 1991 a sequel appeared in the shape of *Political Opposition in the Early Turkish Republic.* From 1977 to 1997 Zürcher worked at the University of Nijmegen, first as a lecturer in Turkish, later as a senior lecturer in History of the Middle East. From 1989 to 1999 he headed the Turkish department of the International Institute of Social History in Amsterdam. From 1993 to 1997 he held an endowed chair of Turkish History in the University of Amsterdam, and since 1997 he has held the chair of Turkish Studies in the University of Leiden. His main research interest is the political and social history of the late Ottoman Empire and the early Turkish Republic. His *Turkey. A Modern History* (first published in 1993) is widely used as textbook in universities in the USA, the UK, the Netherlands and Turkey. It has been translated into Dutch, Turkish, Greek, Arabic, Indonesian and Hebrew. Monographs and textbooks apart, he has edited or co-edited five collections on the social and political history of the late Ottoman Empire and modern Turkey. He has been actively involved in consultations on Turkey's accession to the European Union. In June 2005 he was awarded the Medal of High Distinction of the Ministry of Foreign Affairs of Turkey.

"Multiple Europes"

Multiple Europes is a series that aims to describe and analyse, in a historical perspective, the variety of ways in which social community and images of cohesion have been constructed in Europe. Particular emphasis is given to the idea of "nation," which, without doubt, has been one of the most persistent community categories of the last two hundred years. Various myths, memories, and historical heritages, often contradictory, always shifting, have been mobilised in the processes of community construction, and, taken together, these emphasise the multiplicity of Europe.

It is not only the processes of construction that are multiple – so too are the images of Europe that are produced as a result. Historically, the image of Europe has been both utopian and dystopian. The experience of war has given rise to pacifist dreams, while atomism and disintegration have produced a longing for holism and wholeness. Indeed, the tension between war and peace, atomism and holism, is one of the most obvious elements in the construction of community in Europe. The continent's self-images have oscillated between pride and shame, between, for example, Europe as civilisation and Europe as degeneracy, and these self-images have emerged through various demarcations between Us and the Other. A crucial aspect of such demarcations is the views they project of minorities and immigrants, either as refugees from political regimes and war or as economic migrants.

Another crucial question addressed in the *Multiple Europes* series is how, since the French Revolution, with its rousing call for *liberté, égalité, fraternité*, images of social justice have been used in the construction of community, and how, in political processes, efforts have been made to minimise the evident tension between concepts like freedom and equality.

In this series, the analysis focuses on the levels of both the political, economic and intellectual elites, and the everyday, and considers these levels not only as reflected constructions, but also as unreflected practice. It is in this space between construction and practice that Europe, in its multiple forms, emerges.

Series Editor: Professor **Bo Stråth**
Academy of Finland Distinguished Professor in Nordic,
European and World History at the University of Helsinki

Series Titles

Imagining Europe. Europe and European Civilisation as Seen from its Margins and by the Rest of the World, in the Nineteenth and Twentieth Centuries, Michael WINTLE (ed.), No.42, 2008

Figures of Authority. Contributions towards a Cultural History of Governance from the Seventeenth to the Twentieth Century, Peter BECKER & Rüdiger VON KROSIGK (eds.), No.41, 2008

Europe and the Historical Legacies in the Balkans, Raymond DETREZ & Barbara SEGAERT (eds.), No.40, 2008

Histoire économique et sociale de la construction européenne, René LEBOUTTE, n° 39, 2008

Collective Traumas. Memories of War and Conflict in 20th-Century Europe, Conny MITHANDER, John SUNDHOLM & Maria HOLMGREN TROY (eds.), No.38, 2007

Reflections on Europe. Defining a Political Order in Time and Space, Hans-Åke PERSSON and Bo STRÅTH (eds.), No.37, 2007

London – Berlin. Authenticity, Modernity, and the Metropolis in Urban Travel Writing from 1851 to 1939, Hagen SCHULZ-FORBERG, No.36, 2006

(Un)Doing Europe. Discourses and Practices of Negotiating the EU Constitution, Michał KRZYŻANOWSKI & Florian OBERHUBER, No.35, 2007

Developing Cultural Identity in the Balkans. Convergence vs. Divergence, Raymond DETREZ & Pieter PLAS (eds.), No.34, 2005

Statehood Before and Beyond Ethnicity. Minor States in Northern and Eastern Europe, 1600-2000, Linas ERIKSONAS & Leos MÜLLER (eds.), No.33, 2005

Berlin en musée. Changements sociopolitiques et usages du patrimoine, Véronique CHARLÉTY, n° 32, 2005

Democratisation in Turkey. The Role of Political Parties, Huri TÜRSAN, No.31, 2004

Unravelling Civilisation. European Travel and Travel Writing, Hagen SCHULZ-FORBERG, No.30, 2005

Nation States and Minority Rights. A Constitutional Law Analysis (provisional title), Christina JOHNSSON, No.29, forthcoming

Living Boundaries. Frontiers and Identity in the Basque Country, Zoe BRAY, No.28, 2004

La Serbie du prince Miloš à Milošević, Yves TOMIĆ, n° 27, 2003, second printing 2004

National Heroes and National Identities. Scotland, Norway and Lithuania, Linas ERIKSONAS, No.26, 2004

Constructions identitaires et dynamiques politiques, Lucy BAUGNET, n° 25, 2003

Identity and Justice. Conflicts, Contradictions and Contingencies, Debbie RODAN, No.24, 2004

Homelands. Poetic Power and the Politics of Space, Ron ROBIN & Bo STRÅTH (eds.), No.23, 2003

Figures d'Europe/Images and Myths of Europe, Luisa PASSERINI (ed.), No.22, 2003

European Discourses on Rights. The Quest for Statehood in Europe: The Case of Slovakia, Lubica UČNÍK, No.21, 2003

La survie libanaise ou l'expérience de la différence. Un nécessaire élan vers l'harmonie multiconfessionnelle, Christophe DUBOIS, n° 20, 2002

Gendering Europeanisation, Ulrike LIEBERT (ed.) with Stefanie SIFFT, No.19, 2003

Multilingualism in Europe. A Case Study, THE MADARIAGA EUROPEAN FOUNDATION, Ed. by Lorna CARSON, No.18, 2003, second printing 2005

Narration, Navigation, and Colonialism. A Critical Account of Seventeenth- and Eighteenth-Century English Narratives of Adventure and Captivity, Jamal Eddine BENHAYOUN, No.17, 2007.

A Topography of Memory. Representations of the Holocaust at Dachau and Buchenwald in Comparison with Auschwitz, Yad Vashem and Washington, DC, Isabelle ENGELHARDT, No.16, 2002

L'Orient représenté. Charles Montagu Doughty et les voyageurs anglais du XIXe siècle, Janice DELEDALLE-RHODES, n° 15, 2000

Between Europe and Islam. Shaping Modernity in a Transcultural Space, Almut HÖFERT & Armando SALVATORE (eds.), No.14, 2000, second printing 2004

Across the Atlantic. Cultural Exchanges between Europe and the United States, Luisa PASSERINI (ed.), No.13, 2000

Migrations et migrants dans une perspective historique. Permanences et innovations – Migrations and Migrants in Historical Perspective. Permanencies and Innovations, René LEBOUTTE (ed.), No.12, 2000

Re-Imagining the Nation. Debates on Immigrants, Identities and Memories, Mette ZØLNER, No.11, 2000

Europe and the Other and Europe as the Other, Bo STRÅTH (ed.), No.10, 2000, third printing 2004

Myth and Memory in the Construction of Community. Historical Patterns in Europe and Beyond, Bo STRÅTH (ed.), No.9, 2000

Civitas Europa. *Cities, Urban Systems and Cultural Regions between Diversity and Convergence*, Léonce BEKEMANS & Eduard MIRA (eds.), No.8, 2000

European Societies between Diversity and Convergence (Vol. II), Léonce BEKEMANS & Robert PICHT (eds.), No.7, 1996

Questions de minorités en Europe, Panayotis GRIGORIOU (ed.), n° 5, 1994

L'identité européenne. Analyses et propositions pour le renforcement d'une Europe pluraliste, Robert PICHT (dir.), n° 4, 1993

European Societies between Diversity and Convergence (Vol. I), Léonce BEKEMANS & Robert PICHT (eds.), No.3, 1993

Portable Roots. Voices of Expatriate Wives, Enid GORDON & Morwenna JONES, No.2, 1990

Migrants et travail en Europe, René KAHN, n° 1, 1989